This report contains the collective views of
ternational group of experts and does not nece
represent the decisions or the stated policy oɪ ᴛʜᴇ
United Nations Environment Programme, the International Labour Organisation, or the World Health
Organization.

I0049349

Environmental Health Criteria 145

METHYL PARATHION

First draft prepared by Dr R.F. Hertel and co-workers,
Fraunhofer Institute of Toxicology and Aerosol
Research, Hanover, Germany

Published under the joint sponsorship of
the United Nations Environment Programme,
the International Labour Organisation,
and the World Health Organization

World Health Organization
Geneva, 1993

The **International Programme on Chemical Safety (IPCS)** is a joint venture of the United Nations Environment Programme, the International Labour Organisation, and the World Health Organization. The main objective of the IPCS is to carry out and disseminate evaluations of the effects of chemicals on human health and the quality of the environment. Supporting activities include the development of epidemiological, experimental laboratory, and risk-assessment methods that could produce internationally comparable results, and the development of manpower in the field of toxicology. Other activities carried out by the IPCS include the development of know-how for coping with chemical accidents, coordination of laboratory testing and epidemiological studies, and promotion of research on the mechanisms of the biological action of chemicals.

WHO Library Cataloguing in Publication Data

Methyl parathion.

(Environmental health criteria ; 145)

1.Environmental exposure 2. Methyl parathion - adverse effects
3. Methyl parathion - poisoning 4.Methyl parathion - toxicity
I.Series

ISBN 92 4 157145 4   (NLM Classification: WA 240)
ISSN 0250-863X

The World Health Organization welcomes requests for permission to reproduce or translate its publications, in part or in full. Applications and enquiries should be addressed to the Office of Publications, World Health Organization, Geneva, Switzerland, which will be glad to provide the latest information on any changes made to the text, plans for new editions, and reprints and translations already available.

The designations employed and the presentation of the material in this publication do not imply the expression of any opinion whatsoever on the part of the Secretariat of the World Health Organization concerning the legal status of any country, territory, city or area or of its authorities, or concerning the delimitation of its frontiers or boundaries.

The mention of specific companies or of certain manufacturers' products does not imply that they are endorsed or recommended by the World Health Organization in preference to others of a similar nature that are not mentioned. Errors and omissions excepted, the names of proprietary products are distinguished by initial capital letters.

CONTENTS

ENVIRONMENTAL HEALTH CRITERIA FOR METHYL PARATHION

WHO TASK GROUP ON ENVIRONMENTAL HEALTH CRITERIA FOR METHYL PARATHION

Members

Dr L.A. Albert, Consultores Ambientales Asociados, S.C., Xalapa, Veracruz, Mexico (*Vice-Chairman*)

Dr S. Dobson, Ecotoxicology and Pollution Section, Institute of Terrestrial Ecology, Monks Wood Experimental Station, Abbots Ripton, Huntingdon, Cambridgeshire, United Kingdom

Dr D.J. Ecobichon, Pharmacology and Therapeutics, McGill University, Montreal, Canada (*Chairman*)

Dr R.F. Hertel, Fraunhofer Institute of Toxicology & Aerosol Research, Hanover, Germany (*Co-rapporteur*)

Dr S.K. Kashyap, National Institute of Occupational Health, Meghaninagar, Ahmedabad, India

Dr I. Nordgren, Department of Toxicology, Karolinska Institute, Stockholm, Sweden

Dr K.C. Swentzel, Toxicology Branch II, Health Effects Division, US Environmental Protection Agency, Washington, DC, USA (*Co-rapporteur*)

Dr M. Tasheva, Department of Toxicology, Institute of Hygiene and Occupational Health, Medical Academy, Sofia, Bulgaria

Dr L. Varnagy, Department of Agrochemical Hygiene, University of Agricultural Sciences, Institute for Plant Protection, Keszthely, Hungary

Observers

Dr W. Flucke, Bayer AG, Fachbereich Toxikologie, Institut für Toxikologie Landwirtschaft, Wuppertal, Germany

Secretariat

Dr K.W. Jager, International Programme on Chemical Safety, World Health Organization, Geneva, Switzerland (*Secretary*)

Dr E. Matos, Unit of Carcinogen Identification and Evaluation, International Agency for Research on Cancer (IARC), Lyon, France

NOTE TO READERS OF THE CRITERIA MONOGRAPHS

Every effort has been made to present information in the criteria monographs as accurately as possible without unduly delaying their publication. In the interest of all users of the environmental health criteria monographs, readers are kindly requested to communicate any errors that may have occurred to the Director of the International Programme on Chemical Safety, World Health Organization, Geneva, Switzerland, in order that they may be included in corrigenda.

* * *

A detailed data profile and a legal file can be obtained from the International Register of Potentially Toxic Chemicals, Palais des Nations, 1211 Geneva 10, Switzerland (Telephone No. 7988400 - 7985850).

NOTE: The proprietary information contained in this monograph cannot replace documentation for registration purposes, because the latter has to be closely linked to the source, the manufacturing route, and the purity/impurities of the substance to be registered. The data should be used in accordance with paragraphs 82-84 and recommendations paragraph 90 of the Second FAO Government Consultation (1982).

ENVIRONMENTAL HEALTH CRITERIA FOR METHYL PARATHION

A WHO Task Group on Environmental Health Criteria for Methyl Parathion met at the World Health Organization, Geneva from 19 to 23 August 1991. Dr K.W. Jager, IPCS, welcomed the participants on behalf of Dr M. Mercier, Director of the IPCS, and the three IPCS cooperating organizations (UNEP/ILO/WHO). The Group reviewed and revised the draft and made an evaluation of the risks for human health and the environment from exposure to methyl parathion.

The first draft of the EHC on methyl parathion was prepared by Dr R.F. Hertel and his co-workers of the Fraunhofer Institute of Toxicology and Aerosol Research in Hanover, Germany. The same group assisted in the preparation of the second draft, incorporating comments received following circulation of the first drafts to the IPCS contact points for Environmental Health Criteria monographs.

Dr K.W. Jager of the IPCS Central Unit was responsible for the scientific content of the monograph, and Mrs M.O. Head of Oxford for the editing.

The efforts of all who helped in the preparation and finalization of the monograph are gratefully acknowledged.

1. SUMMARY AND EVALUATION, CONCLUSIONS, RECOMMENDATIONS

1.1 Summary and evaluation

1.1.1 Exposure

Methyl parathion is an organophosphorus insecticide that was first synthesized in the 1940s. It is relatively insoluble in water, poorly soluble in petroleum ether and mineral oils, and readily soluble in most organic solvents. Pure methyl parathion consists of white crystals; technical methyl parathion is a light tan colour with a garlic-like odour. It is thermally unstable and undergoes fast decomposition above pH 8.

Gas chromatography, with either alkali flame ionization (AFID) or flame photometric (FPD) detectors, is the most common method for the determination of methyl parathion. Detection limits range from 0.01 to 0.1 μg/litre in water, and from 0.1 to 1 ng/m^3 in air. HPLC and TLC are also useful methods of detection.

The distribution of methyl parathion in air, water, soil, and organisms in the environment is influenced by several physical, chemical, and biological factors.

Studies using model ecosystems and mathematical modelling indicate that methyl parathion partitions mainly into the air and soil in the environment with lesser amounts going to plants and animals. There is virtually no movement through soil and neither the parent compound nor its breakdown products will normally reach groundwater. Methyl parathion in air mainly arises from the spraying of the compound, though some volatilization occurs with the evaporation of water from leaves and the soil surface. Background atmospheric levels of methyl parathion in agricultural areas range from not detectable to about 70 ng/m^3. Air concentrations after spraying have been shown to decline rapidly over 3 days reaching background levels after about 9 days. Levels in river water (in laboratory studies) declined to 80% of the initial concentration after 1 h and 10% after 1 week. Methyl parathion is retained longer in soil than in air or water, though retention is greatly influenced by soil type; sandy soil can lose residues of the compound more rapidly than loams. Residues on plant surfaces and within leaves decline rapidly with half lives of the order of a few hours; complete loss of methyl parathion occurs within about 6-7 days.

Animals can degrade methyl parathion and eliminate the degradation products within a very short time. This is slower in lower vertebrates and invertebrates than in mammals and birds. Bioconcentration factors are low and the accumulated methyl parathion levels transitory.

By far the most important route for the environmental degradation of methyl parathion is microbial degradation. Loss of the compound in the field and in model ecosystems is more rapid than that predicted from laboratory studies. This is because of the variety of microorganisms capable of degrading the compound in different habitats and circumstances. The presence of sediment or plant surfaces, which increases the microbial populations, increases the rate of breakdown of methyl parathion.

Methyl parathion can undergo oxidative degradation, to the less stable methyl paraoxon, by ultraviolet radiation (UVR) or sunlight; sprayed films degrade under UVR with a half-life of about 40 h. However, the contribution of photolysis to total loss in an aquatic system has been estimated to be only 4%. Hydrolysis of methyl parathion also occurs and is more rapid under alkaline conditions. High salinity also favours hydrolysis of the compound. Half-lives of a few minutes were recorded in strongly reducing sediments, though methyl parathion is more stable when sorbed on other sediments.

In towns in the centre of agricultural areas of the USA, methyl parathion concentrations in air varied with season and peaked in August or September; maximum levels in surveys were mainly in the range of 100-800 ng/m^3 during the growing season. Concentrations in natural waters of agricultural areas in the USA ranged up to 0.46 μg/litre, with highest levels in summer. There are only small numbers of published reports on residues of methyl parathion in food throughout the world. In the USA, residues of methyl parathion in food have generally been reported at very low levels with few individual samples exceeding maximum residue limits (MRLs). Only trace residue levels of methyl parathion were detected in the total dietary studies reported. Methyl parathion residues were highest in leafy (up to 2 mg/kg) and root (up to 1 mg/kg) vegetables in market basket surveys in the USA between 1966 and 1969. Food preparation, cooking, and storage all cause decomposition of methyl parathion residues further reducing exposure of humans. Raw vegetables and fruits may contain higher residues after misuse.

The production, formulation, handling, and use of methyl parathion as an insecticide are the principal potential sources of exposure of humans. Skin contact and, to a lesser degree, inhalation are the main routes of exposure of workers.

In a study on farm spray-men (with unprotected workers using ultra-low-volume (ULV) handsprays) an intake of 0.4-13 mg of methyl parathion per 24 h was calculated from the excreted *p*-nitrophenol in the urine. Early re-entry into treated crops is a further source of exposure.

The general population may be exposed to air-, water-, and food-borne residues of methyl parathion as a consequence of agricultural or forestry practices, the misuse of the agent resulting in the contamination of fields, crops, water, and air through off-target spraying.

1.1.2 Uptake, metabolism, and excretion

Methyl parathion is readily absorbed via all routes of exposure (oral, dermal, inhalation) and is rapidly distributed to the tissues of the body. Maximum concentrations in various organs were detected 1-2 h after treatment. Conversion of methyl parathion to methyl paraoxon occurs within minutes of administration. A mean terminal half-life of 7.2 h was determined in dogs following intravenous (i.v.) administration of methyl parathion. The liver is the primary organ of metabolism and detoxification. Methyl parathion or methyl paraoxon are mainly detoxified in the liver through oxidation, hydrolysis, and demethylation or dearylation with reduced glutathione (GSH). The reaction products are *O*-methyl *O-p*-nitrophenyl phosphorothioate or dimethyl phosphorothioic or dimethylphosphoric acids and *p*-nitrophenol. Therefore, it is possible to estimate exposure by measuring the urinary excretion of *p*-nitrophenol; urinary excretion of *p*-nitrophenol by human volunteers was 60% within 4 h and approximately 100% within 24 h. The metabolism of methyl parathion is important for species selective toxicity, and the development of resistance. The elimination of methyl parathion and metabolic products occurs primarily via the urine. Studies conducted on mice with radiolabelled (^{32}P-methyl parathion) revealed 75% of radioactivity in the urine and up to 10% radioactivity in the faeces after 72 h.

1.1.3 Effects on organisms in the environment

Microorganisms can use methyl parathion as a carbon source and studies on a natural community showed that concentrations of up to 5 mg/litre increased biomass and reproductive activity. Bacteria and actinomycetes showed a positive effect of methyl parathion while fungi and yeasts were less able to utilize the compound. A 50% inhibition of growth of a diatom occurred at about 5 mg/litre. Cell growth of unicellular green algae was reduced by between 25 and 80 μg methyl parathion/litre. Populations of algae became tolerant after exposure for several weeks.

Methyl parathion is highly toxic for aquatic invertebrates with most LC_{50}s ranging from < 1 μg to about 40 μg/litre. A few arthropod species are less susceptible. The no-effect level for the water flea (*Daphnia magna*) is 1.2 μg/litre. Molluscs are much less susceptible with LC_{50}s ranging between 12 and 25 mg/litre.

Most fish species in both fresh and sea water have LC_{50}s of between 6 and 25 mg/litre with a few species substantially more or less sensitive to methyl parathion. The acute toxicity for amphibians is similar to that for fish.

Population effects have been seen on communities of aquatic invertebrates in experimental ponds treated with methyl parathion. The concentrations needed to cause these effects would occur only with overspraying of water bodies and, even then, would last for only a short time. Population effects are, therefore, unlikely to be seen in the field. Kills of aquatic invertebrates would be unlikely to lead to lasting effects.

Care should be taken to avoid overspraying of ponds, rivers, and lakes, when using methyl parathion. The compound should never be sprayed under windy conditions.

Methyl parathion is a non-selective insecticide that kills beneficial species as readily as pests. Kills of bees have been reported following spraying of methyl parathion. Incidents concerning bees were more severe with methyl parathion than with other insecticides. Africanized honey bees are more tolerant of methyl parathion than European strains.

Methyl parathion was moderately toxic for birds in laboratory studies, with acute oral LD_{50}s ranging between 3 and 8 mg/kg body weight. Dietary LC_{50}s ranged from 70 to 680 mg/kg diet. There is no indication that birds would be adversely affected from recommended usage in the field.

Extreme care must be taken to time methyl parathion spraying to avoid adverse effects on honey bees.

1.1.4 Effects on experimental animals and in vitro test systems

Oral LD_{50} values of methyl parathion in rodents range from 3 to 35 mg/kg body weight, and dermal LD_{50} values, from 44 to 67 mg/kg body weight.

Methyl parathion poisoning causes the usual organophosphate cholinergic signs attributed to accumulation of acetylcholine at nerve endings. Methyl parathion becomes toxic when it is metabolized to methyl paraoxon. This conversion is very rapid. No indications of organophosphorous-induced, delayed neuropathy (OPIDN) have been observed.

Technical methyl parathion was found not to have any primary eye or skin irritating potential.

In short-term toxicity studies, using various routes of administration on the rat, dog, and rabbit, inhibition of plasma, red blood cell, and brain ChE, and related cholinergic signs were observed. In a 12-week feeding study on dogs, the no-observed-adverse-effect level (NOAEL) was 5 mg/kg diet (equivalent to 0.1 mg/kg body weight per day). In a 3-week dermal toxicity study on rabbits, the no-observed-effect-level (NOEL) was 10 mg/kg body weight daily. Inhalation exposure for 3 weeks indicated a NOEL of 0.9 mg/m^3 air. At 2.6 mg/m^3, only slight inhibition of plasma ChE was observed.

Long-term toxicity/carcinogenicity studies were carried out on mice and rats. The NOEL for rats was 0.1 mg/kg body weight per day, based on ChE inhibition. There is no evidence of carcinogenicity in mice and rats, following long-term exposure. In another 2-year study on rats, however, there was evidence of a peripheral neurotoxic effect at a dose of 50 mg/kg diet.

Methyl parathion has been reported to have DNA-alkylating properties *in vitro*. The results of most of the *in vitro* genotoxicity studies on both bacterial and mammalian cells were positive, while 6 *in vivo* studies using 3 different test systems produced equivocal results.

In reproduction studies, at toxic dose levels (ChE inhibition), there were no consistent effects on litter size, number of litters, pup survival rates, and lactation performance. No primary teratogenic or embryotoxic effects were noted.

1.1.5 Effects on human beings

Several cases of acute methyl parathion poisoning have been reported. Signs and symptoms are those characteristic of systemic poisoning by cholinesterase-inhibiting organophosphorous compounds. They include peripheral and central cholinergic nervous system manifestations appearing as rapidly as a few minutes after exposure. In case of dermal exposure, symptoms may increase in severity for more than one day and may last several days.

Studies on volunteers, following repeated, long-term exposures, suggest that there is a decrease in blood cholinesterase activities without clinical manifestations.

No cases of organophosphorous-induced, delayed peripheral neuropathy (OPIDN) have been reported. Neuro-psychiatric sequelae have been reported in cases of multiple exposure to pesticides including methyl parathion.

An increase in chromosomal aberrations has been reported in cases of acute intoxications.

No human data were available to evaluate the teratogenic and reproductive effects of methyl parathion.

The available epidemiological studies deal with multiple exposure to pesticides and it is not possible to evaluate the effects of long-term exposure to methyl parathion.

1.2 Conclusions

Methyl parathion is a highly toxic organophosphorus ester insecticide. Overexposure from handling during manufacture, use, and/or accidental or intentional ingestion may cause severe or fatal poisoning. Methyl parathion formulations may, or may not, be irritating to the eyes or to the skin, but are readily absorbed. As a consequence, hazardous exposures may occur without warning.

Methyl parathion is not persistent in the environment. It is not bioconcentrated and is not transferred through food-chains. It is degraded rapidly by many microorganisms and other forms of wild-life. This insecticide is likely to cause damage to ecosystems only in instances of heavy over-exposure resulting from misuse or accidental spills; however, pollinators and other beneficial insects are at risk from spraying with methyl parathion.

Exposure of the general population to methyl parathion residues occurs predominantly via food. If good agricultural practices are followed, the Acceptable Daily Intake (0-0.02 mg/kg body weight), established by FAO/WHO, will not be exceeded. Dermal exposure may also occur through accidental contact with foliar residues in sprayed fields or in areas adjacent to spraying operations as a consequence of off-target loss of the chemical.

With good work practices, hygienic measures, and safety precautions, methyl parathion is unlikely to present a hazard for those occupationally exposed.

1.3 Recommendations

● For the health and welfare of workers and the general population, the handling and application of methyl parathion should be entrusted only to competently supervised and well-trained applicators, who must follow adequate safety measures and use the chemical according to good application practices.

● The manufacture, formulation, agricultural use, and disposal of methyl parathion should be carefully managed to minimize contamination of the environment.

● Regularly exposed workers should receive appropriate monitoring and health evaluation.

● To minimize risks for all individuals, a 48-h interval between the spraying and re-entry into any sprayed area is recommended.

- Pre-harvest intervals should be established and enforced by national authorities.

- In view of the high toxicity of methyl parathion, this agent should not be considered for use in hand-applied, ULV spraying practices.

- Do not overspray water bodies. Choose spraying times to avoid killing pollinating insects.

- Information on the health status of workers exposed only to methyl parathion (i.e., in manufacture, formulation) should be published, in order to better evaluate the risks of this chemical for human health.

- More definitive studies should be conducted on residues of methyl parathion in fresh foods.

- A more definitive genotoxic assessment of methyl parathion should be conducted.

2. IDENTITY, PHYSICAL AND CHEMICAL PROPERTIES, ANALYTICAL METHODS

2.1 Identity

2.1.1 Primary constituent

Molecular formula: $C_8H_{10}NO_5PS$

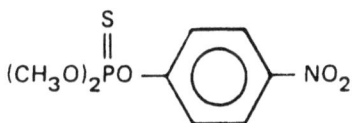

$$(CH_3O)_2\overset{\overset{\displaystyle S}{\|}}{P}O - \langle \bigcirc \rangle - NO_2$$

Relative molecular mass: 263.23

Common names: methyl parathion
accepted by
ESA (Entomological Society of America)
JMAF (Japanese Ministry of Agriculture, Fisheries and Food)
WHO (World Health Organization)

parathion-methyl
accepted by
BSI (British Standards Institution)
ISO (International Organization for Standardization)

metaphos
accepted by the USSR

CAS chemical name: *O,O*-dimethyl *O*-(4-nitro-phenyl)-phosphorothioate

IUPAC systematic name: *O,O*-Dimethyl *O*-4-nitrophenylphos-phorothioate

CAS registry number: 298-00-0

RTECS number: TG 0175000

EINECS number: 206-050-1

EEC number: 015-035-00-7

Common synonyms:

> Demethylfenitrothion; dimethyl *para*-nitrophenyl monothiophosphate; *O,O*-dimethyl O-(*para*-nitrophenyl) phosphorothioate; dimethyl *para*-nitrophenyl phosphorothionate; dimethyl 4-nitrophenyl phosphorothionate; *O,O*-dimethyl O-(*para*-nitrophenyl) thionophosphate; dimethyl *para*-nitrophenyl thiophosphate; *O,O*-dimethyl-*O*-(*para*-nitrophenyl) thiophosphate; dimethyl parathion; ENT 17292; metaphos; methyl-parathion; methylthiophos; MPT; NCI CO2971; parathion methyl homolog; phosphorothioic acid *O,O*-dimethyl O-(4-nitrophenyl) ester; phosphorothioic acid *O,O*-dimethyl O-(*para*-nitrophenyl) ester BAY 11405; 8056 HC; E601

2.1.2 Technical product

Major trade names:

> A-Gro; Azofos; Azophos; Bladan M; Cekumethion; Dalf; Divithion; Drexel Methyl Parathion 4E & 601; Dygun; Dypar; Ekatox; Folidol M, M40 & 80; Fosferno M50; Gearphos; Mepaton; Meptox; Metacid 50; Metacide; Metafos; Metaphos; Methyl-E 605; Methyl Fosferno; Methylthiophos; Metron; M-Parathion; Niletar; Niran M-4; Nitran; Nitrox; Nitrox 80; Oleovofotox; Parapest M-50; Parataf; Paratox; Paridol; Parton M; Penncap M & MLS; Sinafid M-48; Sixty-Three Special E.C. Insecticide; Tekwaisa; Thiophenit; Thylpar M-50; Toll; Unidol; Vertac Methyl Parathion; technical product 80%, Wofatox; Wolfatox.

.2.1. Purity

Technical methyl parathion is available as a solution containing 80% active ingredient (a.i.), 16.7% xylene, and 3.3% inert ingredients.

The following impurities were identified in one sample of technical-grade methyl parathion: *O,O*-dimethyl-*S*-methyl dithiophosphate, nitroanisol, nitro-phenol, isomers of methyl parathion, and the dithio-analogue of methyl parathion (Warner, 1975).

2.2 Physical and chemical properties

Physical state: pure: white crystalline solid or powder (National Fire Protection Association, 1986)

technical (80%) pure: light to dark tan liquid (Worthing & Walker, 1987)

Melting point: 37-38 °C (The Merck Index, 1983)
35-36 °C (Worthing, 1983)

Freezing point: about 29 °C (technical product) (Worthing & Walker, 1987)

Density/specific gravity:
1.358 at 20 °C/40 °C (d_4^{20} 1.358) (The Merck Index, 1983)

Vapour pressure: 1.3 mPa at 20 °C (Worthing & Walker, 1987)

Octanol/water partition coefficient:
$\log K_{ow}$ = 2.68 (measured)
$\log K_{ow}$ = 1.81-3.43 (reported range) (Hansch & Leo, 1987)

Water solubility: 55-60 mg/litre at 25 °C (pure)
(Midwest Research Institute, 1975;
National Research Council, 1977)
37.7 mg/litre at 19 °C (pure)
(Bowman & Sans, 1979)
57 mg/litre at 22 °C (anal. grade)
(Sanders & Seiber, 1983)

Nonaqueous solubility: soluble in ethanol, chloroform,
aliphatic solvents, and slightly
soluble in light petroleum

Volatility (pure): 0.14 mg/m^3 at 20 °C (Spencer, 1982)

Odour: like rotten eggs or garlic (technical grade)
(Midwest Research Institute, 1975; Anon.,
1984)

Odour threshold: 0.0125 mg/m^3 (Akhmedov, 1968)

Other properties: hydrolyses and isomerizes easily (White-
Stevens, 1971)

Half-life in aqueous solution at 20 °C, pH 1-5:
175 days (Melnikov, 1971)

2.3 Conversion factors

1 ppm methyl parathion = 10.76 mg/m^3 at 25 °C, 1066 mbar

1 mg methyl parathion/m^3 = 0.0929 ppm

2.4 Analytical methods

2.4.1 Sampling, extraction, clean-up

Standardized methods for the determination of various residues
are reported in the *Manual of pesticide residue analysis* (Thier &
Zeumer, 1987).

4.1.1   *Plant material (tobacco, fruits, vegetables, crops with low oil (fat) content)*

(*a*) Extraction

Three extraction methods have mainly been used, all of which are suitable for multiresidue analysis.

(1) Soxhlet extraction with chloroform - 10% methanol has been proposed for field-weathered crops by Bowman (1981).

(2) Acetonitrile combined with various amounts of water has been used by Mills et al. (1963), Wessel (1967), Osadchuk et al. (1971), Luke et al. (1975), and Stahr et al. (1979). The plant material is homogenized in a blender with acetonitrile, in some instances after the addition of Celite (Nelson, 1967; Funch, 1981;). High-moisture products (fruits and vegetables) are extracted with pure acetonitrile while samples of dry products (hays, grains, feedstuff) are blended with acetonitrile-water (65:35). Extraction is followed by solvent partitioning into petroleum ether with the addition of sodium chloride (Mills et al., 1963; Wessel, 1967; Nelson, 1967) into dichloromethane (Funch, 1981), and dichloromethane/hexane (10:200) (Osadchuk et al., 1971).

(3) Acetone was preferred as the solvent in particular in multiresidue analysis by Becker (1971), Pflugmacher & Ebing (1974), Sagredos & Eckert (1976), Becker (1979), Specht & Tillkes (1980), Miellet (1982), Sonobe et al. (1982), Luke & Doose (1983), Luke & Doose (1984), Ebing (1985), Andersson & Ohlin (1986), Vogelsang & Thier (1986), Gyorfi et al. (1987), Thier & Zeumer (1987), and Becker & Schug, (1990). In some instances, celite was added. Depending on the water content of the sample, water was added. In a second step, the acetone extracts were further extracted with either dichloromethane, dichloro-methane/petroleum ether, or dichloromethane/*n*-hexane. The extract was dried over anhydrous sodium sulfate, reduced in volume in a Kuderna-Danish concentrator, and subjected to further clean-up.

Extraction with acetone-*o*-xylene (19:1) (Ross & Harvey, 1981), toluene/hexane (75:25) (Johansson, 1978), chloroform (Ault et al., 1979), or supercritical fluid extraction using methanol (Capriel et al., 1986), has also been reported.

(*b*) Column clean-up

The published clean-up procedures are usually suitable for multiresidue analysis. For plant material with a low fat content, 3 column clean-up procedures have been developed.

(1) The oldest method involves the use of chromatography on Florisil (often topped with anhydrous sodium sulfate) (Mills et al., 1963; Nelson, 1967; Schnorbus & Phillips, 1967; Wessel, 1967; Beckman & Garber, 1969; Osadchuk et al., 1971; Luke et al., 1975; Johansson, 1978; Gretch & Rosen, 1984, 1987). Although it has been claimed that organophosphorous pesticides are partially lost during Florisil clean-up (Luke et al., 1975), high recoveries (usually > 80 %) have been reported for methyl parathion. Various solvents and solvent mixtures are used for chromatography on Florisil including: diethylether/petroleum ether, ethylether/hexane, and acetone/toluene, diethylether/petroleum ether being the most frequently used. Fractionation is achieved by increasing successively the diethylether content. Florisil clean-up is usually used for a combined clean-up of organochlorine and organophosphorous pesticides. Luke et al. (1975) reported that gas chromatography (GC) with a thermionic detector was sufficiently selective to detect organophosphorous pesticides without Florisil clean-up.

(2) Alternatively, clean-up of pesticides in multiresidue analysis has been achieved by chromatography on charcoal (Becker, 1971, 1979; Miellet, 1982; Sonobe et al., 1982; Luke & Doose, 1984; Ebing, 1985; Gyorfi et al., 1987). To this end, charcoal is mixed with silica gel (1:15) (and sometimes also celite or magnesia). In most instances, elution is achieved with mixtures of dichloromethane/acetone/toluene (e.g., 5:1:1) (Ebing, 1985; Thier & Zeumer, 1987). Recoveries are high (often > 90 %). Charcoal clean-up is particularly suited for dry products (< 10 % water). The

simultaneous clean-up of organochlorine and organophosphorous pesticides is also possible with chromatography on charcoal.

(3) In recent years, a clean-up of pesticides in multiresidue analysis by gel permeation chromatography (GPC) has become popular (Pflugmacher & Ebing, 1974; Ault et al., 1979; Specht & Tillkes, 1980; Andersson & Ohlin, 1986; Vogelgesang & Thier, 1986; Steinwandter, 1988). The stationary phase consists, in most instances, of Bio Beads SX3 (a polystyrene gel). Ethyl acetate/cyclohexane (1:1), dichloromethane/cyclohexane (1:1) and, more recently, acetone/cyclohexane (3:1) have been used as elution mixtures. Gel permeation chromatography is mainly used to protect the GC column and the GC detector against contamination. GPC removes material of higher relative molecular mass. Recoveries > 85% have been reported. Frequently, GPC is combined with the additional purification step of silica gel chromatography (Specht & Tillkes, 1980; Andersson & Ohlin, 1986; Vogelsaifng & Thier, 1986) where elution is achieved with toluene/hexane (35:65), followed by toluene and acetone/toluene, with increasing acetone content. However, while the additional clean-up by silica gel column chromatography is important when organochlorine pesticides are present, it is not necessary for organophosphorous pesticides if analysis is performed by gas chromatography with flame photometric detection.

4.1.2 Dairy products, products with a high fat content (edible fats)

Clean-up techniques for products with a high fat content have been reviewed by Waters (1990). Florisil column chromatography and gel permeation chromatography are also suited for a clean-up of samples with a high fat content. In addition, clean-up using normal phase HPLC has been reported (Gillespie & Waters, 1986). Fat is dissolved in *n*-hexane and fractionated on silica gel HPLC using dichloromethane/hexane as solvent. However, complete separation of methyl parathion from the fat is not achieved. As an alternative, fat is adsorbed on aluminum oxide (Luke & Doose, 1984) or on Calflo E (calcium silicate) (Specht, 1978; Thier & Zeumer, 1987). Finally, a sweep codistillation clean-up of edible oils has been reported by

Storherr et al. (1967) and Watts & Storherr (1967). This method has been standardized also for plant material (Thier & Zeumer, 1987). After extraction of the sample with ethyl acetate, the concentrated extract is injected into a heated glass column packed with glass wool or glass beads followed by the injection of ethyl acetate or petroleum ether in a nitrogen stream. The nitrogen carrier gas sweeps the volatile component through the tube to a condensing bath and through an Arnakrom scrubber tube to a collection tube. Sweep codistillation may be followed by a further Florisil clean-up.

The extraction and clean-up of vegetable oil can be speeded up by performing extraction and clean-up in one step using a system of three ready-to-use cartridges in series (Extralut-3, Sep-Pack silicadel and Sep-Pack C_{18}) where the assembled columns are eluted with acetonitril (saturated with *n*-hexane) (Di Muccio et al., 1990).

2.4.1.3 Blood, body fluids

Methyl parathion is extracted from blood with hexane or benzene and analysed without further clean-up (Gabica et al., 1971; De Potter et al., 1978). No extraction is necessary if methyl parathion is determined by polarography (Zietek, 1976).

Measurement of the urinary metabolites and the cholinesterase activity were used to supervise the exposure of workers coming into contact with methyl parathion or parathion and to observe their elimination in cases of poisoning (see section 5.3) (Elliot et al., 1960; Arterberry et al., 1961; Shafic & Enos, 1969; Wolfe et al., 1970; Ware et al., 1974b; NIOSH, 1976).

2.4.1.4 Soil, sediments

Methyl parathion is extracted from soil with acetone, acetone/*n*-hexane or hexane/isopropanol (Schutzmann et al., 1971; Agishev et al., 1977; Garrido & Monteoliva, 1981; Wegman et al., 1984; Kjoelholt, 1985). It is partitioned in a second step into dichloromethane. While several authors determine the pesticides without further clean-up, additional silica gel adsorption chromatography has been used by Wegman et al., (1984) and Kjoelholt (1985). The recovery of methyl parathion is 70-85%.

When sediments are analysed, elemental sulfur represents a particular problem. Kjoelholt et al. separated the sulfur by tetrabutylammonium hydrogensulfate (Kjoelholt, 1985), while Schutzmann et al. (1971) refluxed the sediment extract with Raney copper.

For the extraction, the sediment mixed with sand and sodium sulfate can be placed into a column and eluted using acetone : dichloromethane (1:1) (Belisle & Swineford, 1988).

4.1.5 Water

Extraction and concentration of methyl parathion from water is achieved either by liquid/liquid extraction (Kawahara et al., 1967; Pionke et al., 1968; Mestres et al., 1969; Konrad et al., 1969; Zweig & Devine, 1969; Schutzmann et al., 1971; Coburn & Chau, 1974; Chmil et al., 1978; Chernyak & Oradovskii, 1980; Miller et al., 1981; Spingarn et al., 1982; Bruchet et al., 1984; Albanis et al., 1986; Li & Wang, 1987; Brodesser & Schoeler, 1987), or by adsorption on polymeric material (Paschal et al., 1977; Le Bel et al., 1979; Agostiano et al., 1983; Xue, 1984; Clark et al., 1985). Various solvents have been used for solvent extractions including: diethyl ether/hexane (1:1), benzene, petroleum ether, hexane/isopropanol; chloroform, dichloromethane, and ethyl acetate. Recoveries have been high (in most instances > 90 %). If the liquid/liquid extraction is scaled up using a "Goulden large sample extractor" and 120 litre of water, detection limits may be lower by a factor of about 150 compared with 1-litre samples (i.e., a detection limit of 2.5 ng/litre (ppt) has been achieved for methyl parathion) (Foster & Rogerson, 1990). The extraction efficiency can be further improved by continuous liquid-liquid extraction, which allows the use of non-polar solvents as n-pentane (Bruchet et al., 1984; Brodesser & Schoeler, 1987). Water samples are frequently analysed for pesticides without further clean-up, while Florisil clean-up has been used in some instances (Mestres et al., 1969; Miller et al., 1981).

High concentration factors are achieved, if methyl parathion (and other pesticides) are adsorbed on polymeric material, such as XAD-2 (Paschal et al., 1977; Le Bel et al., 1979), XAD-4 (Xue et al., 1984), Tenax (Agostiano et al., 1983) or Porapack Q (Clark et al., 1985). Elution from XAD is achieved with diethyl ether, acetone/hexane (15:85), diethyl ether-hexane (85:15). Recoveries are >90 %. If Tenax is used, both solvent elution (diethyl ether) or thermoelution can be used to desorb the pesticides. Solid-phase extraction (using C-18 cartridges) will become the method of choice for the rapid extraction of organophosphorous insecticides from water (Swineford & Belisle, 1989; Sherma & Bretschneider, 1990).

2.4.1.6 Air

Most methods for the determination of pesticides in air have been developed as multiresidue methods. Pesticides in air are either absorbed in liquids or adsorbed on polymeric material. Thus, pesticides may be trapped in ethylene glycol, which is subsequently extracted with dichloromethane (Tessari & Spencer, 1971; Sherma & Shafik, 1975) or they may be trapped on glass beads coated with cottonseed oil (Compton, 1973). Further clean-up is achieved by silica gel or Florisil column chromatography.

Among the solid polymeric material used to trap pesticides, polyurethane foam (PUF) is by far the most popular (Lewis et al., 1977; Rice et al., 1977; Lewis & McLeod, 1982; Lewis & Jackson, 1982; Belashova et al., 1983; Beine, 1987). Air can be collected both with low-volume ($\approx$ 4 litre/min) or high-volume samplers (up to 250 litre/min). PUF can be reused after careful cleaning (e.g., with 5% diethyl ether in *n*-hexane). In some instances, Tenax, Chromosorb 102, or Porapack R is sandwiched between PUF plugs to enhance the collection efficiency. Collection efficiencies in excess of 80% have been reported for methyl parathion. A filter may be added to remove particulate matter (Lewis et al., 1977). Methyl parathion is usually determined without further clean-up. Finally, XAD-4 (Wehner et al., 1984) and silica gel (Klisenko & Girenko, 1980; Liang & Zhang, 1986) have been used as solid trapping materials.

2.4.1.7 Formulations

When analysing formulations, the determination of by-products and impurities is an important objective. A variety of instrumental techniques have been used for the analysis of formulations including: gas chromatography (Jackson, 1976; Jackson, 1977a), high performance liquid chromatography (Jackson, 1977b), infrared analysis (Goza, 1972), P-31-nuclear magnetic resonance spectroscopy (Greenhalgh et al., 1983), and spectrophotometry after alkaline hydrolysis to *p*-nitrophenol (Blanco & Sanchez, 1989). An interlaboratory study has been carried out using both GC (Jackson, 1977a) and HPLC (Jackson, 1977b). With both methods, coefficients of variation of 1.7% have been determined. The instrumental techniques are described below.

2.4.2 *Instrumental analytical methods*

4.2.1 *Gas chromatography*

Gas chromatrophic (GC) methods for the determination of pesticides (including methyl parathion) have been reviewed by Ebing (1987).

Organophosphorous pesticides, including methyl parathion, are sufficiently volatile and thermally stable to be amenable to gas chromatography and it is by far the most important method for the determination of methyl parathion. This technique provides the good resolution necessary for multiresidue analysis. Moreover, very sensitive and specific detectors are available, in particular for the analysis of organophosphorous pesticides.

(*a*) Detectors

The two most widely used detectors for organophosphorous pesticides are the alkali flame ionization detector (AFID) and variations of this detector (thermionic detector (Patterson, 1982), nitrogen-phosphorous detector) and the flame photometric detector (FPD) (Bowman, 1981). The AFID makes use of the phenomenon that the flame ionization detector yields enhanced response to nitrogen- and phosphorus-containing compounds, in the presence of alkali metal salts. The detection limit is in the low picogram range. The detector discriminates against other compounds 30-50 fold. The flame photometric detector (FPD) operates with a cool, hydrogen rich flame for the detection of phosphorus- and sulfur-containing compounds, which form POH and S_2 species. These species emit light at 526 nm (POH) and 394 nm (S_2), which is monitored by using interference filters and a photomultiplier. The detector is easy to operate and results are reproducible. The detector is highly specific. The response of 100 ng of parathion is 130 000 times greater than that of an equal amount of aldrin. Furthermore, It is of advantage that any solvent can be used with the detector. For the determination of methyl parathion the P mode is the method of choice, though the S mode can also be used (sensitivity 10 times lower) as methyl parathion contains both P and S atoms.

Finally, the electron capture detector (ECD) is sometimes used for the analysis of methyl parathion as it responds not only to the $P=S$ moiety, but in particular to the NO_2 group.

(*b*) Columns

A definite identification of a pesticide by its retention time on one column is not possible. Analysis on at least one further column with a stationary phase of different polarity is necessary to confirm the identity of a compound.

Packed columns are frequently used for pesticide residue analysis, though resolution is substantially poorer compared with capillary columns and identification of the pesticides is less specific. Solid supports are usually of the Chromosorb W type. In some instances, Gaschrom Q has also been used. A large variety of stationary phases, used either alone or in admixture, have been employed. The most frequently used phases are DC 200, QF-1, OV 17, OV-101, OV-210, and SE-30. Relative retention times for many stationary phases have been reported by several authors for a large variety of pesticides (up to 600 compounds including other industrial chemicals) (Bowman & Beroza, 1967; Ambrus et al., 1981b; Daldrup et al., 1981; Prinsloo & de Beer, 1987; Saxton, 1987; Suprock & Vinopal, 1987; Omura et al., 1990).

Packed column GC allows the separation of only a limited number of pesticides. Capillary columns exhibit a considerably better separation efficiency than packed columns. Such capillary columns have been used by several authors for methyl parathion analysis (Krijgsman & van den Kamp, 1976; Ripley & Braun, 1983; Stan & Goebel, 1983; Ebing, 1985; Andersson & Ohlin, 1986; Vogelsang & Thier, 1986). Retention time data on a SE-30 capillary column have been reported (Ripley & Braun, 1983). Several injection techniques for capillary columns have been compared (Stan & Goebel, 1984; Stan & Mueller, 1988). Cold splitless (PTV) injection appears to be best suited for organophosphorous pesticide analysis. The resolution can be further improved by applying two-dimensional capillary gas chromatography using two columns of different polarity (Stan & Mrowetz, 1983).

4.2.2 High performance liquid chromatography (HPLC)

The main advantage of HPLC is its ability to analyse compounds that are heat labile, such as phenylurea and carbamates. As stated above, organophosphorous pesticides including methyl parathion are sufficiently heat stable for analysis using gas chromatography and there is no direct need to use HPLC. Thus, relatively few studies dealing with the HPLC analysis of methyl parathion have been reported.

HPLC analysis has been achieved using reversed phase chromatography, with acetonitrile/water (60:40) (Funch, 1981), or methanol/acetic acid/water (32:0.6: 47.4) as solvents, and UV-detection (Zhao & Wang, 1984). HPLC conditions for 166 pesticides including methyl parathion were reported by Lawrence & Turton (1978). Retention data of 560 pesticides and other industrial chemicals have been published by Daldrup et al. (1981, 1982) using two gradient systems.

Sharma et al. (1990) developed a method for the rapid quantitative analysis of organophosphorus (including methyl parathion) and carbamate pesticides using HPLC and refractive index detection.

HPLC appears to be particularly suited for the analysis of polar metabolites of methyl parathion (Abe et al., 1979).

Fluorogenic labelling of organophosphorous pesticides leads to an improvement in sensitivity. Such labelling can be achieved by hydrolysis of the compounds to the corresponding phenols and derivatization with dansyl chloride (5-dimethylamino-naphthalene-1-sulfonyl chloride) (Lawrence et al., 1976). Besides the UV and fluorescence detector, electrochemical detectors have been used for the detection of methyl parathion using amperometric detection in the reductive mode (Bratin et al., 1981; Clark et al., 1985) or polaro-graphic detection (Koen & Huber, 1970). Acetonitrile/water with additional acetate buffer is used as solvent. The response is similar to the UV detector, but there is less interference from the plant material (Clark et al., 1985).

2.4.2.3 *Thin layer chromatography (TLC)*

Thin layer chromatography is well suited for the analysis of organophosphorous pesticides, even if it is not as specific as GC (Kawahara et al., 1967; Schütz & Schindler, 1974; Thielemann, 1974; Katkar & Barve, 1976; Lawrence et al., 1976; Curini et al., 1980; Daldrup et al., 1981; Pfeiffer & Stahr, 1982; Korsos & Lantos, 1984). Usually, silica gel G plates are used with a variety of solvent or solvent mixtures. These include benzene, chloroform/ cyclohexane, *n*-hexane/acetone, chloroform/benzene, dichloro- methane/acetone. Silver nitrate is frequently used as spray reagent, which, in the presence of organophosphorous pesticides, leads to white spots against a black background (Pfeiffer & Stahr, 1982).

As an alternative, an enzymatic reaction has been frequently applied to detect organophosphorous compounds on TLC plates (Mueller, 1973; Leshev & Talanov, 1977; Ambrus et al., 1981a; Bhaskar & Kumar, 1981; Devi et al., 1982). This method makes use of the fact that cholinesterase (from horse serum or cow liver) hydrolises 1-naphthyl acetate to 1-naphthol, which reacts either with Fast Blue Salt B or *p*-nitrobenzenediazoniumfluoroborate to form a coloured complex. If methyl parathion is inhibiting the enzyme reaction, white spots on a red or orange background appear. The sensitivity may be enhanced if methyl parathion is oxidized to methylparaoxon by reaction with bromine or hydrogen peroxide.

2.4.2.4 *Spectrophotometry*

Colorimetric methods, which were of importance during the early years of organophosphorous pesticide analysis, have largely been replaced by chromatographic methods.

The inhibition of cholinesterase by organophosphorous pesticides, described above, is also the basis of a photometric method (Archer & Zweig, 1959; Kumar & Ramasundari, 1980; Bhaskar & Kumar, 1982, 1984; Kumar, 1985). Sadar et al. (1970) made use of the fact that cholinesterase hydrolyses the non fluorescent *N*-methyl- indoxylacetate to the highly fluorescent indoxyl. This reaction is again inhibited by methyl parathion.

In another spectrophotometric method, methyl parathion is treated with hydroxylamine hydrochloride and sodium nitroprusside, under alkaline conditions, to form a water-soluble, coloured complex (Sastry & Vijaya, 1986). The method is rapid and accurate and can be used for formulations and for residues in fruits and vegetables.

4.2.5 Polarography

Polarography and various modifications of this method, i.e., the "differential pulse polarography" (DPP), have been used repeatedly to determine methyl parathion and other organophosphorous compounds with a nitro group (Nangniot, 1966; Gajan, 1969; Kheifets et al., 1976; Zietek, 1976; Smyth & Osteryoung, 1978; Kheifets et al., 1980; Khan, 1988; Reddy & Reddy, 1989). The method allows the simultanous determination of parathion, methyl parathion, paraoxon, EPN, and the metabolite 4-nitrophenol (Zietek, 1976) in blood, without prior extraction. Polarography has been proposed as confirmatory method for the determination of methyl parathion (and three further pesticides). A collaborative study of 10 laboratories showed a coeffient of variation of 15-16% (Gajan, 1969). In addition the method was applied to water analysis (Kheifets et al., 1976, 1980; Bourquet et al., 1989). Bourquet et al. (1989) showed a 20-50 increase in sensitivity when "adsorptive stripping voltametry" was used instead of DPP.

4.2.6 Mass spectrometry

Coupled gas chromatography/electron impact mass spectrometry (GC/MS) is a particularly valuable method for confirming pesticide residues in various environmental samples. Methyl parathion shows an abundant $m/z = 109$, 125, and 263 ($M^{+\cdot}$) under electron impact conditions (Mestres et al., 1977; Wilkins, 1990). Under positive ion chemical ionization mass spectrometry (methane), the protonatic molecule is the most abundant ion (m/z 264) while the structure specific fragment at m/z 125 is due to $(CH_3O)_2$ $P=S^+$ (8.8%) (Holmstead & Casida, 1974). The negative ion chemical ionization spectrum shows the typical thiophenolate fragment at $m/z = 154$ ($^-S-C_6H_4-NO_2$) (Nielsen, 1985).

In addition, field ionization (FI) and field desorption (FD) mass spectrometry have been applied repeatedly in the determination of of methyl parathion (Damico et al., 1969; Klisenko et al., 1981; Schulten & Sun, 1981; Golovatyi et al., 1982). The FD spectra show little fragmentation and, thus, are not well suited for environmental analysis. Among the newer mass spectrometric techniques, tandem mass spectrometry (MS/MS) shows more promise for organophosphorous pesticide analysis, as this technique enhances the selectivity of the method and thus may reduce the necessary clean-up. Under MS/MS conditions (chemical ionization), the protonated molecule forms an abundant fragment at m/z 125 ($(CH_3O)_2 P=S^+$) (Hummel & Yost, 1986; Roach & Andrzejewski, 1987).

HPLC/MS of methyl parathion has been demonstrated (De Wit et al., 1987; Betowski & Jones, 1988; Farran et al., 1990). As this method is more difficult to handle and less sensitive and reproducible than GC/MS, there is no need to use it in routine analysis, except when other thermally labile pesticides are to be determined together with organophosphorous compounds.

2.4.3 Detection limits

Detection limits are rarely reported. When plant material was analysed, the detection limit for the overall method (extraction, clean-up, analysis) was 10-100 µg/kg when gas chromatography with AFID or FPD was used. In water analysis, substantially better detection limits were achieved (usually 0.01-0.1 µg/litre), which may be further reduced if a large-scale extractor is used (Foster & Rogerson, 1990). In air analysis, detection limits have been reported to be 0.1-1 ng/m^3.

2.4.4 Confirmatory method

A confirmatory derivatization method was proposed by Lee et al. (1984). Following hydrolysis with KOH, 4-nitrophenol was derivatized with pentafluoro benzyl bromide to the corresponding ether. Analysis is carried out by GC with ECD. Levels as low as 0.01 ppb can be confirmed.

Table 1. Sampling, extraction, clean-up, and determination of methyl parathion[a]

Matrix	Sampling, extraction, clean-up	Analytical method	Recovery (%)	Detection limit[b] (µg/kg or litre)	References
fruits, vegetables	extr.: acetonitrile, part.: petroleum ether, clean-up: Florisil	GC (ECD, TID) TLC	86-92	n.r.	Wessel (1967)
plant material, dairy products	extr.: propylene carbonate, clean-up: Florisil	GC (ECD, TID)	82-95	n.r.	Schnorbus & Phillips (1967)
fruits, vegetables, fat, oil	extr.: acetonitrile, part.: dichloromethane + hexane, clean-up: Florisil	GC (ECD)	90-98	n.r.	Osadchuck et al. (1971)
vegetables	extr.: acetone, part.: dichloromethane/petroleum ether, clean-up: Florisil	GC (ECD, TID)	93 (celery)	n.r.	Luke et al. (1975)
apples	extr.: toluene + n-hexane, clean-up: Florisil	GC (ECD)	93	1-20	Johansson (1978)

Table 1 (continued)

Matrix	Sampling, extraction, clean-up	Analytical method	Recovery (%)	Detection limit[b] (µg/kg or litre)	References
vegetables	autom. extraction + clean-up: Florisil	n.r.	91-104 (pepper)	n.r.	Gretch & Rosen (1984)
food	extr.: acetone, part.: dichloromethane, clean-up: GPC + silica gel	GC	n.r.	n.r.	Specht & Tillkes (1980)
fruits, vegetables	extr.: acetone, part.: dichloromethane hexane, clean-up: GPC and silica gel	GC (ECD,FPD, TID)	> 80	10-100	Andersson & Ohlin (1986)
vegetables, fruits, crops	extr.: trichloromethane, clean-up: GPC	GC (FPD)	93-105	n.r.	Ault et al. (1979)

Table 1 (continued)

vegetables	extr.: acetone, part.: dichloromethane, clean-up: GPC	GC (TID)	85-95	n.r.	Pflugmacher & Ebing (1974)
-	clean-up: GPC	n.r.	n.r.	n.r.	Steinwandter (1988)
-	clean-up: cellulose column	n.r.	82	n.r.	Stahr et al. (1979)
fruits, vegetables	extr.: acetonitrile, part.: dichloromethane	HPLC (UV 280)	77-87	10	Funch (1981)
honey bees, beeswax, pollen	extr.: acetone o-xylene	GC (FPD)	92-101	1	Ross & Harvey (1981)
plants, soil	extr.: supercritical methanol	GC (ECD, AFID)	38	n.r.	Capriel et al. (1986)
tobacco	extr.: hexane/acetone, clean-up: alumina	GC (FPD)	99-104	20	Sagredos & Eckert (1976)
vegetables	extr.: acetone, part.: dichloromethane, clean-up: charcoal	GC (ECD,TID, FPD)	n.r.	n.r.	Gyorfi et al. (1987)

Table 1 (continued)

Matrix	Sampling, extraction, clean-up	Analytical method	Recovery (%)	Detection limit[b] (μg/kg or litre)	References
plant material	extr.: acetone, part.: dichloromethane	GC (AFID, ECD)	92-103	n.r.	Becker (1971)
plant material	extr.: acetone, part.: dichloromethane, clean-up: charcoal	GC (ECD, AFID)	92-103	n.r.	Becker (1979)
plant material	extr.: acetone, clean-up: charcoal/Florisil	HPLC	n.r.	n.r.	Miellet (1982)
barley, malt, hops	extr.: acetone or acetonitrile, part.: hexane, clean-up: charcoal	GC (FPD)	82	30	Sonobe et al. (1982)

Table 1 (continued)

low moisture products (pepper)	extr.: acetone, part.: dichloromethane/petrol, ether, clean-up: charcoal	GC (FPD)	93	n.r.	Luke & Doose (1983)
ready-to-eat foods	extr.: acetone part.: dichloromethane, clean-up: + GPC silica gel	GC (ECD, TID)	n.r.	0.7-1.8	Vogelsang & Thier (1986)
honey bees	extr.: acetone clean-up: charcoal	GC (ECD)	91	15	Ebing (1985)
milk, oilseeds	fat adsorbed on alumina extr.: acetonitrile, part.: petroleum ether	GC (ECD, FPD)	n.r.	80	Luke & Doose (1984)
fat	ad.: of fat on Calflo E		n.r.	n.r.	Specht (1978)
edible oils	sweep co-distillation	GC (TID)	95	10 (mg/kg)	Storherr et al. (1967)

Table 1 (continued)

Matrix	Sampling, extraction, clean-up	Analytical method	Recovery (%)	Detection limit[b] (µg/kg or litre)	References
edible oils	extr.: petroleum ether, clean-up: HPLC	GC(FPD)	83-107	n.r.	Gillespie & Walters (1989)
milk	sweep co-distillation	GC (TID)	> 87	n.r.	Watts & Storherr (1967)
blood	extr.: n-hexane	GC (FPD)	n.r.	3	Gabica et al. (1971)
serum	extr.: benzene	GC (AFID)	69	2	De Potter et al. (1978)
blood	no extr.	polarography		7×10^{-8} mol	Zietek (1976)
soil	extr.: acetone/hexane	GC (TID)	n.r.	n.r.	Agishev et al. (1977)
soil	extr.: acetone/hexane	TLC (silica gel)	n.r.	n.r.	Garrido & Monteoliva (1981)
soil, sediment	extr.: acetone/hexane, clean-up: ad. chrom.	GC (AFID)	71	0.17	Kjoelholt (1985)

Table 1 (continued)

soil	extr.: acetone, part.: dichloromethane, clean-up: silica gel	GC (TID)	78-85	5	Wegman et al. (1984)
soil, water, sediment	extr.: hexane/isopropanol, desulfurization with Raney copper	GC (ECD)	45	n.r.	Schutzmann et al. (1971)
water	diethylether/hexane or benzene/n-C$_6$, clean-up: TLC	GC (ECD)	n.r.	n.r.	Kawahara et al. (1967)
water	extr.: benzene	GC (TID)	95	n.r.	Pionke et al. (1968)
water	extr.: benzene	GC	92-101	0.001 (?)	Konrad et al. (1969)
water	extr.: petroleum ether	GC	98	0.04	Zweig & Devine (1969)
water	extr.: trichloromethane	TLC	60-95	1	Chmil et al. (1978)

Table 1 (continued)

Matrix	Sampling, extraction, clean-up	Analytical method	Recovery (%)	Detection limit[b] (µg/kg or litre)	References
water	extr.: trichloromethane	GC(TID)	n.r.	0.01	Chernyak & Ora-dovskii (1980)
water/waste-water	extr.: at pH 11: dichloromethane; at pH 2: dichloromethane	GC/MS	60-85	5	Spingarn et al. (1982)
water	extr.: dichloromethane/hexane, clean-up: Florisil	GC (ECD)	n.r.	n.r.	Albanis et al. (1986)
water	extr.: ethylacetate	GC (FPD)	85-91	0.08 ng(abs.)	Li & Wang (1987)
wastewater	extr.: dichloromethane, clean-up: Florisil	GC (FPD)	90	0.75	Miller et al. (1981)
water	extr.: petroleum ether, clean-up: Florisil	GC (ECD)	n.r.	0.5	Mestres et al. (1969)

Table 1 (continued)

water	extr.: dichloromethane (continuous) liquid-liquid)	GC/MS	75	n.r.	Bruchet et al. (1984)
water	extr.: n-pentane (continous liquid-liquid)	GC (TID)	90	0.01	Brodesser & Schoeler (1987)
water	hydrolysis KOH, derivat. penta-fluoro-benzylbromide, clean-up: silica gel	GC (ECD)	95	0.1	Coburn & Chau (1974)
water	ad.: on Tenax, thermoelution	GC (FID/ECD)	62	0.01	Agostiano et al. (1983)
water, run-off water	ad.: XAD-2, elut.: diethylether	HPLC (rev. phase, UV)	99	2	Paschal et al. (1977)
water, drinking-water	ad.: XAD-2, elut.: acetone/hexane	GC (TID, FID)	93-100	15 pg(abs.)	Le Bel et al. (1979)
water	ad.: XAD-4, elut.: diethylether/hexane	GC	n.r.	n.r.	Xue (1984)

Table 1 (continued)

Matrix	Sampling, extraction, clean-up	Analytical method	Recovery (%)	Detection limit[b] (µg/kg or litre)	References
water	ad.: Porapack Q, elut.: acetonitrile	HPLC (rev. phase electro-chem.)	96-105	< 1	Clark et al. (1985)
water	ad.: C-18, elut.: ethyl acetate	TLC	n.r.	0.2 ng(abs.)	Sherma & Bretschneider (1990)
water	ad.: C-18, acetone	GC (FPD)	> 79	n.r.	Swineford & Belisle (1989)
water	extr.: dichloromethane (large-scale extractor)	GC/MS	48	0.0025	Foster & Rogerson (1990)
air	ab.: ethylene-glycol, extr.: dichloromethane, clean-up: silica gel	GC (FPD)	87-97	n.r.	Sherma & Shafik (1975)

Table 1 (continued)

air	ab.: cotton seed oil coated glass beads, clean-up: Florisil	GC (FPD)	91	0.04 ng/m^3	Compton (1973)
air	clothscreen with ethylene glycol, extr.: acetone/hexane, clean-up: alumina + Florisil	GC (ECD/FPD)	93	n.r.	Tessari & Spencer (1971)
air	ad.: silica gel, activated charcoal	GC (ECD/FPD)	n.r.	1 ng (abs.)	Klisenko & Girenko (1980)
air	ad.: silica gel	GC (FPD)	101-104	30 pg (abs.)	Liang & Zhang (1986)
air	ad.: XAD-4, elut: ethylacetate, clean-up: HPLC	GC (ECD, TID)	74	1-3 ng/m^3	Wehner et al. (1984)
air	ad.: PUF, elut: petroleum ether	GC (ECD)	100	n.r.	Rice et al. (1977)

Table 1 (continued)

Matrix	Sampling, extraction, clean-up	Analytical method	Recovery (%)	Detection limit[b] (μg/kg or litre)	References
air	ad.: PUF (high volume sampler)	GC (ECD, FPD)	86	0.1 ng/m^3	Lewis et al. (1977)
air	ad.: PUF (low volume sampler), elut.: diethylether/hexane	GC (ECD, FPD)	80	20 ng/m^3	Lewis & MacLeod (1982)
air	ad.: PUF/other polymers (high volume sampler)	GC	72-91	n.r.	Lewis & Jackson (1982)
air	ad.: PUF, elut.: trichloromethane or acetaldehyde	n.r.	n.r.	n.r.	Belashova et al. (1983)
air	ad.: Tenax, elut.: toluene	GC (FID)	n.r.	2.5 μg/m^3	Beine (1987)

Table 1 (continued)

formulations		GC or HPLC	-	Jackson (1976)
formulations		GC	-	Jackson (1977a)
formulations		HPLC	-	Jackson (1977b)
formulations		IR	-	Goza (1972)
formulations		P-31 NMR	-	Greenhalgh et al. (1983)
formulations	hydrolysis to p-nitrophenol	Spectr.	-	Blanco & Sanchez (1989)

[a] Abbreviations: GC = gas chromatography, TLC = thin-layer chromatography, GPC = gel permeation chromatography, MS = mass spectrometry, HPLC = high performance liquid chromatography, NMR = nuclear magnetic resonance, IR = infrared spectroscopy, ECD = electron capture detector, FID = flame ionization detector, AFID = alkali flame ionization detector, FPD = flame photometric detector, TID = thermionic detector, UV = ultraviolet detector, spectr. = spectrophotometry, extr. = extraction, part. = partitioning, ad. = adsorption, ab. = absorption, elut. = elution, n.r. = not reported, (abs.) = absolute.

[b] μg/kg or litre unless stated otherwise.

Table 2. Methods used in the determination of methyl parathion

Method	Detection limit	Remarks	References
HPLC (UV)	n.r.	analysis of metabolism	Abe et al. (1979)
HPLC (UV) (rev. phase, methanol/acetic acid)	n.r.	in mixtures	Zhao & Wang (1984)
HPLC	n.r.	review on HPLC methods	Lawrence & Turton (1978)
HPLC (fluorescence)	10-20 μg (abs.)-	deriv. with dansyl chloride	Lawrence et al. (1976)
HPLC 1. acetonitrile 2 acetonitrile/phosphoric acid KH_2PO_4/H_2O	n.r.	retention times of 560 compounds	Daldrup et al. (1982)
HPLC 1. acetonitrile 2 acetonitrile/phosphoric acid KH_2PO_4/H_2O	n.r.	retention times of 570 compounds	Daldrup et al. (1981)
HPLC (rev. phase, acetonitrile/H_2O)	10 μg/kg	fruits and vegetables	Funch (1981)

Table 2 (continued)

Method	Detection limit	Detection	Reference
HPLC (rev. phase, acetonitrile/0.01 KCl 0.03 M potassium acetate/H_2O)	1 μg/kg	reduction ampero-metric detection (vegetables, water)	Clark et al. (1985)
HPLC (rev. phase, acetonitrile/sodium acetate/H_2O)	n.r.	• electrochemical detection	Bratin et al. (1981)
HPLC rev. phase (H_2O ethyl alcohol/acetic acid/NaOH)	30 μg/kg	polarographic detection	Koen & Huber (1970)
GC	< 2 ng	TID	Patterson (1982)
GC	n.r.	retention times of 570 compounds	Daldrup et al. (1981)
GC (TID)	20 μg/kg	retention times	Ambrus et al. (1981a,b)
GC	n.r.	retention times of 600 compounds	Saxton (1987)
GC	n.r.	retention times of 42 pesticides	Prinsloo & de Beer (1987)
GC	n.r.	retention times of 78 pesticides	Suprock & Vinopal (1987)

Table 2 (continued)

Method	Detection limit	Remarks	References
GC	n.r.	retentions times of 20 OP-pesticides (milk, corn silage)	Bowman & Beroza (1967)
GC	n.r.	two dimensional GC	Stan & Mrowetz (1983)
GC (FPD)	100 pg	capillary columns, relative retention times	Krijgsman & Van de Kamp (1976)
GC (ECD, TID)	n.r.	capillary columns, simultaneous detection of ECD, TID	Stan & Goebel (1983)
GC	n.r.	retention times of 194 pesticides	Ripley & Braun (1983)
GC	< 0.1 ng	relative retention times of 40 pesticides on 11 phases	Omura et al. (1990)

Table 2 (continued)

GC (ECD)	n.r.	hydrolysis of methyl parathion to 4-nitrophenol, derivat. pentafluorobenzylbromide, Clean up: silica gel	Lee et al. (1984)
TLC (silica gel G)	n.r.	detection with GC	Kawahara et al. (1967)
TLC (silica gel)	0.1 μg	4 solvent mixtures, reduct. to amines	Schütz & Schindler (1974)
TLC (silica gel)	0.06-0.6 μg	saponification and reduct. to p-amino-phenol	Thielemann (1974)
TLC (silica gel G)	n.r.	elut.: n-hexane/acetone	Katkar & Barve (1976)
TLC (silica gel)	n.r.	17 solvent systems, spray reagent: $AgNO_3$	Curini et al. (1980)
TLC (silica gel)	n.r.	elut.: 1. methanol/NH_3/H_2O; 2. dichloromethane/acetone	Daldrup et al. (1981)
TLC (silica gel)	n.r.	elut.: n-heptane/acetone	Pfeiffer & Stahr (1982)

Table 2 (continued)

polarography	140 μg/kg	oscillographic polarography, pesticide residues	Nangniot (1966)
polarography	10 μg/kg	single sweep oscillographic polarography, non-fatty foods	Gajan (1969)
polarography	n.r.	differential oscillographic polarography (water)	Kheifets et al. (1976)
polarography	7×10^{-6} mol/litre	methyl parathion and metabolites in blood	Zietek (1976)
polarography	10^{-8} mol/litre	-	Smyth & Osteryoung (1978)
polarography	n.r.	adsorptive stripping	Bourquet et al. (1988)
polarography	n.r.	-	Kahn (1988)
polarography	$3.9 . 10^{-9}$ mol/litre	polarography, diff. pulse polarography, cyclic voltametry	Reddy & Reddy (1989)

Table 2 (continued)

polarography	140 μg/kg	oscillographic polarography, pesticide residues	Nangniot (1966)
polarography	10 μg/kg	single sweep oscillographic polarography, non-fatty foods	Gajan (1969)
polarography	n.r.	differential oscillographic polarography (water)	Kheifets et al. (1976)
polarography	7×10^{-6} mol/litre	methyl parathion and metabolites in blood	Zietek (1976)
polarography	10^{-8} mol/litre	-	Smyth & Osteryoung (1978)
polarography	n.r.	adsorptive stripping	Bourquet et al. (1988)
polarography	n.r.	-	Kahn (1988)
polarography	$3.9.10^{-9}$ mol/litre	polarography, diff. pulse polarography, cyclic voltametry	Reddy & Reddy (1989)

Table 2 (continued)

Method	Detection limit	Remarks	References
differ. chrono-amperometry	n.r.	water	Kheifets et al. (1980)
spectrophotometry	n.r.	enzymatic reaction (cholinesterase, Fast Blue B)	Kumar (1985)
spectrophotometry	n.r.	reduction to amine, formation of a coloured complex	Sastry & Vijaya (1986)
spectrophotometry	n.r.	reaction with 3-methyl-2-benzothiazolinone	Sastry & Vijaya (1987)
spectrophotometry	n.r.	hydrolysis to 4-nitro-phenol	Ramakrishna & Ramachandran (1978)

[a] Abbreviations: GC = gas chromatography, HPLC = high performance liquid chromatography, TLC = thin layer chromatography, ECD = electron capture detector, TID = thermionic detector, FPD = flame photometric detector, UV = ultraviolet detector, elut. = elution, n.r. = not reported, (abs.) = absolute.

3. SOURCES OF HUMAN AND ENVIRONMENTAL EXPOSURE

3.1 Natural occurrence

Natural occurrence of methyl parathion is unlikely.

3.2 Man-made sources

3.2.1 Production process

Methyl parathion is a representative of the highly active insecticides, the thiophosphorus esters, developed in the 1940s by Schrader, a German chemist. Methyl parathion was introduced as a commercial chemical in 1949. It is synthesized by the reaction of O,O-dimethyl phosphoro-chloridothioate with the sodium salt of 4-nitrophenol (Schrader, 1963).

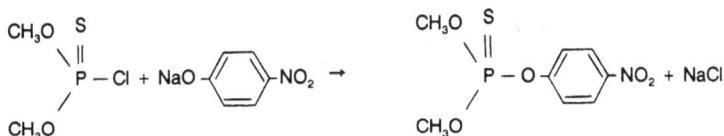

$$CH_3O \backslash \overset{S}{\underset{\parallel}{P}} - Cl + NaO - \langle \rangle - NO_2 \rightarrow CH_3O \backslash \overset{S}{\underset{\parallel}{P}} - O - \langle \rangle - NO_2 + NaCl$$

3.2.2 Loss into the environment

Emissions of methyl parathion during the production process can be disregarded when compared with those from its use as an insecticide. The air emission from a factory in the USA was reported to be around 0.1% of the production level (Archer et al., 1978). The major losses of this insecticide are directly caused by spraying, and evaporation from water surfaces, leaves, and from the soil (Woodrow et al., 1977).

3.2.3 Production

According to the European Directory of Agrochemical Products (1986) and the Directory of World Chemical Producers (1990), methyl parathion is produced throughout the world by many companies. World production in 1966 was 31 700 tonnes, including 14 800 tonnes produced in the USA.

In Table 3, selected countries producing methyl parathion are listed together with their production capacities (Bayer, 1988).

Table 3. Methyl parathion production capacities in different countries[a]

Country	Production capacity in tonnes/year
Brazil	3000
Denmark	15 000
German Democratic Republic	3500
Mexico	8000
India	3000
China	40 000
USSR	5000-10 000

[a] From: Bayer (1988).

3.2.4 World consumption

Recent data from Bayer concerning the consumption of the active ingredient only are reported in Table 4 (Bayer, 1988).

Table 4. Methyl parathion consumption in tonnes in some areas of the world[a]

Region	1984	1985	1986
Africa	191	308	152
North America	2 045	2 776	2 932
South America	9 135	6 555	5 587
Asia, New Zealand, Australia	2 757	3 028	2 620
Western Europe	894	1 087	1 019
Total	15 022	13 754	12 310

[a]From: Bayer (1988).

In 1984, the USA exported 3010 tonnes of methyl parathion (HSDB, 1990).

!.2.5 Formulations

Methyl parathion is used in following formulations:

(1) emulsifiable concentrates (EC) with 19.5%, 40%, 50%, 60% active ingredient (a.i.)
(2) wettable powders containing 40% a.i.
(3) dusts 1.5%, 2%, and 3% methyl parathion,
(4) microencapsulated methyl parathion, and
(5) ready-to-use liquid (less than 1% a.i.).

The usual carriers are: petroleum solvents and clay carriers (such as propargite).

Combinations are available containing parathion, omethoate, tetradifon, prothoate, and petroleum oil.

3.3 Uses

Methyl parathion is a broad-spectrum insecticide with non-systemic contact and stomach action. The normal method of application is foliar spraying by aircraft or ground equipment. Data from 1971 show that most methyl parathion was used for protecting cotton fields (Table 5).

Table 5. Methyl parathion consumption pattern (1971)[a]

Protection of	consumption (%)
cotton	83
soybeans	8
grain including corn	5
wheat	2
tobacco, peanuts, vegetables, and citrus fruits	2

[a]From: HSDB (1990).

Only foliar application of methyl parathion is known. It is used as a contact insecticide and acaricide. There are different routes of application depending on the type of plant to be protected and the organisms killed. The recommended application rate is 0.5-1 kg a.i./ha for vegetables, 1-2 kg/ha for cereals, 1.5-6 kg/ha for fruit trees, 2-5 kg/ha for citrus fruits, and 0.12-1.0 kg/ha for cotton.

4. ENVIRONMENTAL TRANSPORTATION, DISTRIBUTION, AND TRANSFORMATION

4.1 Transportation and distribution between media

The transportation and distribution of methyl parathion in air, water, soil, fauna, and flora are influenced by several physical, chemical, and biological parameters. The transportation and fate of methyl parathion were studied by Gile & Gillett (1981). They used the simulated ecosystem developed at the Corvallis Environmental Research Laboratory of the US EPA (Gillett & Gile, 1976). A 16-h daily light cycle with an average of 27 000 lx at the soil surface was used. The temperatures varied from 18 °C at night to 30 °C during the day. The ecological compartment was ventilated with 10 litre air/min. The simulated ecosystem included alfalfa (*Medicago sativa*) and perennial ryegrass (*Lolium perenne*). Twenty days after planting, different representative kinds of invertebrates (earthworms, nematodes, garden snails) were added to the microcosms. Ten days later, radioactive labelled ^{14}C-methyl parathion (50 μCi) was applied at rates of 0.3, 0.6, and 2.4 kg/ha. One week following the methyl parathion application, a gravid gray-tailed vole (*Microtus canicaudus*) was placed in the model ecosystem. The relative ^{14}C mass balance of the study is shown in the Table 6.

Table 6. ^{14}C mass balance of methyl parathion in a model ecosystem[a]

Samples	Application rate of methyl parathion		
	0.3 kg/ha	0.6 kg/ha	2.4 kg/ha
air	57[b]	46	33
soil	30	30	28
groundwater	0.0	0.1	0.0
plants	12	23	38
animals	1.0	0.6	1.1

[a] From: Gillett & Gile (1976).
[b] %.

Most radioactivity was found in the upper 5 cm of soil. A comparable experiment with *p*-nitrophenol showed a lower soil content and no residues in the groundwater as well.

Crossland & Elgar (1983) used a mathematical model to predict the dispersion and degradation of methyl parathion in freshwater ponds. Basic assumptions of the model were that loss processes could be adequately described in terms of simple partition phenomena and first-order rate kinetics. Predictions of the model were compared with experimentally-obtained data for concentrations of methyl parathion in water and sediment. They started with a concentration of 100 μg methyl parathion/litre pond water. At the limit of the analytical method (0.005 μg/g), they could not find any residues of methyl parathion, 16 days after treatment. The authors described the degradation by a pseudo first order rate constant that was temperature-dependent. Since the degradation of methyl parathion in distilled water (pH not given) was faster than expected and the bacteria concentration was only 10^6/litre, a sediment-catalysed hydrolysis was supposed. Crossland & Bennett (1984) compared degradation of methyl parathion in experimental ponds and laboratory aquaria. Degradation was faster in the natural ponds and faster than predicted from simple mathematical models. Addition of plants, sediment, or sediment with plants, to the laboratory aquaria increased the rate of breakdown of methyl parathion; sediment had the greatest effect reducing half-life from 300 h in water alone to 90-140 h. These findings support the investigation of Goedicke & Winkler (1976), who considered, from their testing of the persistence of different formulations of methyl parathion in soils, that the compound would not contaminate groundwater, if applied at suggested rates and intervals.

4.1.1 Air

Most of this insecticide is directly liberated by spraying. However, a perceptible amount is released simultaneously with evaporation from water surfaces, leaves, or soil (Woodrow et al., 1977).

Air samples were analysed after the application of methyl parathion at a concentration of 1.12 kg/ha (Jackson & Lewis, 1978). The conventional emulsifiable concentrate was compared with an encapsulated formulation. The filter collection efficiency was determined to be 105% and the extraction efficiency was 92%. During the experimental period, the temperature varied from 18 to 34 °C at an average relative humidity of 72%. The results of the analysis of the air samples collected in tobacco-growing areas of North Carolina are shown in Table 7.

Table 7. Concentration of methyl parathion in the air after application[a]

Time (days)	Methyl parathion (mg/m^3)	
	emulsifiable concentrate	encapsulated formulation
0	7.408	3.783
1	3.338	0.330
3	0.584	0.107
6	0.036	0.025
6	0.054	0.019
9	0.013	0.016

[a] From: Jackson & Lewis (1978).

Since the usual atmospheric levels of methyl parathion in the surroundings of agricultural areas range from not detectable to 71 ng/m^3, Jackson & Lewis (1978) discussed the possibility that the concentrations measured on day 9 may have been the result of the background level in the air of the heavily treated areas.

The atmospheric concentration of methyl parathion after spraying in the Kalinin District, Tashkent Province of the Uzbek USSR, during July and August, was determined by Akhmedov (1968). He found that the concentrations measured were dependent on the size of the area of methyl parathion application, the time of application, the temperature, and the wind velocity. In addition, the odour threshold was estimated, and effects on the brain electrical activity, resorption action, dark adaptation, and the light sensitivity of the eyes were studied.

After the aerial treatment of forests, Vrochinsky & Makovsky (1977) measured the following concentrations of methyl parathion in the air (Table 8).

The concentrations of methyl parathion increased in foggy conditions because of the adsorption of the compound on the surface of water aerosols (Goncharuk et al., (1988).

Table 8. Methyl parathion in air after spraying forests[a]

Time (days)	Methyl parathion (mg/m^3)
0	0.12
1	0.05
5	0.024
10	0.0015

[a] From: Vrochinsky & Makovsky (1977).

^{14}C-Methyl parathion was subjected to simulated rainfall (total amount: 2.5, 25, and 38 mm/h) after application of 177 μg ai/cm^2 to an octadecylsilane/trimethylsilane-treated glass slide. The amounts of ^{14}C remaining after washoff were 56%, 6%, and 2% respectively; thus, methyl parathion shows a high rate of washoff (Cohen & Steinmetz, 1986).

4.1.2 Water

Various mechanisms exist for the transportation of methyl parathion following its application to aquatic environments, including: application-associated losses, volatilization, wind erosion, rinsing by rain into groundwater, and transportation as a soil-methyl parathion complex.

Eichelberger & Lichtenberg (1971) estimated the water pollution factor by investigating the persistence of methyl parathion in river water. They used a sealed glass jar containing river water and methyl parathion and applied sunlight and artificial fluorescent light. The initial concentration of methyl parathion was 10 μg/litre (Table 9):

Table 9. Persistence of methyl parathion in river water[a]

Time	% of the initial concentration (10 μg/litre)
1 hour	80[b]
1 week	25
2 weeks	10
4 weeks	0

[a] Adapted from: Eichelberger & Lichtenberg (1971).
[b] Recoveries were rounded off to the nearest 5%.

Badawy & El-Dib (1984) found that methyl parathion was more stable in water of high salinity, such as sea water, than in fresh water.

Because of a collision between two ships in the Mediterranean Sea near Port-Said, Egypt, the sea became contaminated with more than 10 000 kg methyl parathion. Maximum methyl parathion concentrations (96 μlitre/litre) were found 50 m in the drifting direction (surface current, wind). In general, the concentration decreased with distance and time and reached the detection limit up to 80 days after the accident. The residues in sediment gradually increased during the first 20 days (concentration factor 49.5) (Badawy et al., 1984).

Crossland et al. (1986) gave mathematical tools for calculating the fate of chemicals in aquatic systems (because of the importance of the degradation of methyl parathion in water, see also section 4.2).

4.1.3 Soil

Lichtenstein (1975) incorporated an emulsifiable concentration of methyl parathion into the upper 5 inches of a silt loam at a rate of 3.1 mg/kg). One month after treatment, 3.5% of the methyl parathion could be detected in the soil. The author showed that percolating water transported metabolites vertically as well as horizontally. Methyl parathion moved less than 20 cm in a loamy soil following an annual precipitation of 1500 mm (Haque & Freed, 1974).

Bound residues of [ring-^{14}C] methyl parathion in a silt loam were monitored during an incubation period of 49 days (Gerstl & Helling, 1985). After this period, 54% of the initial ^{14}C remained in the soil; of this, 13% was soxhlet-extractable with methanol and 87% was bound residue. Several treatments indicated that bound residues of methyl parathion are not easily released (i.e., converted to an extractable form), but that they are slowly mineralized to CO_2.

A simulated spillage of emulsifiable or microencapsulated formulations of methyl parathion on soil (sandy loam; pH ranging from 6.6 to 7.8, with a mean of 7.2) was studied for 45 months by Butler and coworkers (1981). The uptake of the insecticide was studied in five different experiments. The soil was contaminated with: a) 51% emulsifiable concentrate formulation (E.C.), b) dilute drum rinse of E.C., c) 22% microencapsulated formulation (M.C.), d) dilute drum rinse of M.C., and e) a solid cake of M.C.

microencapsulated formulation of the initial values (Table 10). At 45 months, soil residues of methyl parathion had decreased by 64% for emulsifiable concentrate spills, and 68% for the soil beneath the microencapsulated cake; the residue in the cake itself only decreased by 31% (Table 10). Soil residue concentrations from the simulated drum rinses (Table 10) were very low by 45 months (emulsifiable concentrate) and by one year (microencapsulated formulation).

Performing laboratory experiments, Davidson et al. (1980) showed that, at low application rates (24.5 mg/kg), methyl parathion was non-persistent in soils (Webster & Cecil) but was persistent following application of large quantities (10015 mg/kg). Therefore, it is impossible to predict the behaviour of methyl parathion at high applications rates on the basis of results following low application rates.

Table 10. Persistence of methyl parathion in sandy loam soil and in solid cake material following contamination of the soil with different formulations of methyl parathion[a]

Time (months)	Mean concentrations of Methyl parathion (mg/kg)				
	E.C.[b] (51%)	E.C. (rinse)	M.C.[c] (22%)	M.C. (rinse)	M.C. (cake)
0	48 900	17 600	30 800	2 140	379 000
1	33 700	10 800	14 200	940	258 000
3	25 300	7 000	17 100	550	305 000
12	20 900	3 800	20 000	0.15	87 500
20	20 800	1 400	13 300	230	149 000
45	17 500	130	9 800	n.r.[d]	262 000

[a] Modified from: Butler et al. (1981).
[b] E.C. = emulsifiable concentrate.
[c] M.C. = microencapsulated formulation.
[d] n.r. = not recorded.

4.1.4 Vegetation and wildlife

Residue levels of methyl parathion on foliage depend on the formulation, the method of application, humidity, rain, temperature, dust levels etc. Kido et al. (1975) investigated surface and internal residue levels of methyl parathion on grape leaves treated with methyl parathion sprays (at the rate of 0.84 kg a.i./ha.); 90.2% of the initial surface residue was lost from the leaves one day after application. The major portion, over 60%, of the total residues was found in the internal portion of the leaves, and over 99% of the total residues had been lost, 5 days after application. Overhead sprinkler irrigation of the vines had only a slight, or no, effect on the reduction of methyl parathion residues (Kido et al., 1975). The residual life of methyl parathion on cotton can be extended by application at dusk rather than dawn. For example, methyl parathion decreased to less than 50% after 4 h in sunlight, but only to 84% after the same time at night (Ware et al., 1980). The persistence of methyl parathion following application to cotton was also increased by combining it with molasses (Ware et al., 1980), toxaphene (Buck et al., 1980; Ware et al., 1980; Bigley et al., 1981), camphene (Bigley et al., 1981), or cedar oil (Bigley et al., 1981). Ware et al. (1983) compared surface residues of methyl parathion on cotton foliage. When applied to cotton fields (at 1.1 kg/ha) as a typical, low-volume spray diluted with water versus ultra-low-volume (ULV) application using vegetable oil as the carrier. Forty-eight hours after application as an aqueous dilution, 1.8 % of the initial residue remained compared with 7.2 % after application as ULV. Cole et al. (1986) sprayed methyl parathion 4E (EC) in either water or water-crop oil (6:1) at 8 litres of a 1.8% dilution/ha on a 5 ha plot of cotton using a pawnee airplane. The residues found in the leaves sprayed with the mixture containing crop oil were higher than those in water-sprayed leaves in all samples collected after the treatment (Table 11).

The drift from a commercial aerial application of methyl parathion was quantified by Draper & Street (1981) by determining leaf surface residues of methyl parathion in a treated alfalfa field and an adjoining non-target pasture (with quackgrass, Agropyron repens, as predominant species). Four hours after the pesticide spraying by plane (0.27 kg/litre emulsifiable concentrate; 0.7 litre/ha; in the morning) 2.8 mg methyl parathion/kg were present as foliar residues in the target field, and 0.26 mg/kg, in the untreated non-target pasture. At both places, the foliar residues of the parent compound dissipated rapidly with time.

Table 11. Comparison of methyl parathion residues in cotton leaves treated with water sprays and with water-oil sprays[a]

Days after treatment	Methyl parathion concentration	
	water	water-oil formulation
1	14.80 ± 8.74[b]	27.70 ± 7.99
2	9.17 ± 7.15	9.68 ± 4.29
3	2.30 ± 0.89	7.48 ± 2.85
4	1.52 ± 0.31	8.70 ± 4.58
5	1.96 ± 1.49	5.97 ± 2.61

[a] From: Cole et al. (1986).
[b] mg/kg mean ± SE.

The time-dependent decrease in the residues of 2 different formulations of methyl parathion applied to tobacco plants was evaluated. Methyl parathion in either the emulsifiable or the encapsulated form was applied at rates of 0, 0.56, and 1.12 kg/ha. Samples were collected before spraying and within 10 min of the application. It was observed that the encapsulated formulation of methyl parathion did not decompose as fast as the emulsifiable form (Leidy et al., 1977). Varis (1972) tried to determine the influence of plant growth on the loss of methyl parathion residues in sugar beet seedlings. Methyl parathion was applied as a dust formulation (1.5%) at 20 kg/ha, 14 days after sowing. The residue methyl parathion concentration in the plants decreased to about 50% within 24 h. Within 6 days, the methyl parathion residue was reduced by 90%, 73% reduction being due to plant growth.

Fuhremann & Lichtenstein (1978) performed experiments with unextractable, soil-bound residues of radioactive labelled methyl parathion and measured the potential pick up of the ^{14}C-containing residues. Earthworms (*Lumbricus* spp.) and oat (*Avena sativa* L.) plants were able to release and incorporate some soil-bound, ^{14}C-ring-labelled methyl parathion. Oat plants were found to release more chemical from the soil than the earthworms.

Following applications of insecticides (including methyl parathion) to nearby sugarcane or cotton fields, alterations in brain acetylcholinesterase activity were found in birds living in brushland within the Lower Rio Grande Valley of South Texas (Custer & Mitchell, 1987). These alterations might have resulted from exposure during the use of agricultural fields as feeding or resting sites.

4.1.5 Entry into the food-chain

Methyl parathion hydrolyses faster than parathion. Because of the physical and chemical properties of methyl parathion, its pollution potential seems to be very small. Therefore, the most probable entry into the food-chain seems to be directly via residues on vegetables or crops.

Since animals can degrade methyl parathion and excrete the degradation products within a very short time, a risk from eating meat seems to be unlikely. However, there may be an additional hazard from methyl parathion bound to glucosides (Dorough, 1978).

4.2 Biotransformation

4.2.1 Degradation involving biota

Both field and laboratory studies have been conducted on the degradation of methyl parathion were. Data suggest that biodegradation is the major degradative pathway in eutrophic systems, whereas absorption, photolysis, and hydrolysis are more important in oligotrophic systems.

The half-lives of methyl parathion residues reported in the literature for plants were relatively short, but varied with ambient conditions (see also section 4.1.4).

Singh et al. (1978) recorded half-lives of methyl parathion applied to urd (*Phaseolus mungo* Roxb.) and pea (*Pisum sativum* (L) var. *arvense* Poir.) at the rate of 0.63 and 1.25 kg a.i. per ha, respectively. Half-lives were 1.7 and 2.5 days for urd and 2.0 and 2.7 days for pea, respectively. Foliar residues of methyl parathion on alfalfa treated by aircraft (0.27 kg/litre, emulsifiable concentrate) dissipated showing a first-order half-life of 12 h. This calculation is based on initial slopes of semi-logarithmic plots (Draper & Street, 1981). The authors, however, noted that dissipation kinetics appeared to be greater than first-order. The times required for a 50% reduction in methyl parathion residues in cotton foliage were determined to be 4.4-5.4 h (emulsifiable concentrate) or 28.1 h (encapsulated formulation) following application at a rate of

0.28 kg/ha (Smith et al., 1987). Based on data previously reported by Ware et al. (1974a) following application of methyl parathion to cotton (1.12 kg/ha), half-lives of 12 h (emulsifiable concentrate) and 70 h (encapsulated formulation) were calculated (Smith et al., 1987). In another study using emulsifiable concentrate formulations of methyl parathion at a rate of 1.15 kg/ha, a 50% disappearance time of 2-4 h was calculated for methyl parathion on cotton plants (Willis et al., 1985). A half-life of 0.96 days was described for methyl parathion residues (initial concentration = 0.4 μg/cm) on apple leaf surfaces (Goedicke, 1989).

A single report is available on the persistence of methyl parathion in a submerged aquatic macrophyte (*Hydrilla verticilla*) and a fish (carp), both initially exposed to 3.8 mg methyl parathion/litre. The first order half-lives were 7.9 and 5.4 days, respectively (Sabharwal & Belsare, 1986).

The half-life of methyl parathion in a soil (not characterized in detail) has been reported to be about 45 days (Menzie, 1972). In another study it was calculated to be as short as 2.7 days (Singh et al., 1978), possibly due to the high pH of the soil (pH = 8.6) and temperature (28 °C-33 °C). Half-lives of 12 and 22 days were measured for methyl parathion in 2 soils (pH = 6.1 and 5.5, respectively) when incubated at 22 °C (Möllhoff, 1981). Concentrations of methyl parathion in a loamy sand soil (pH = 5.3) decreased from a level of about 5 mg/kg to 0.3 mg/kg during a period of 57 days (Goedicke & Winkler, 1976). Thirty days following treatment, 3.1% of initial residues of methyl parathion were found in a soil (clay?) of a field treated with 5.6 kg/ha (Lichtenstein & Schulz, 1964) (see also section 4.1.3).

During an incubation study under aerobic conditions, methyl parathion was degraded mainly to CO_2 and 4-nitrophenol, and, to a minor extent, to desmethyl parathion (Möllhoff, 1981). Methyl parathion may be degraded in the environment by: *a*) hydrolysis to *p*-nitrophenol and dimethylthiophosphoric acid; or *b*) nitro-group reduction to methyl aminoparathion (e.g., Sharmila et al., 1988). Hydrolysis can be both chemical and microbial while nitro-group reduction is essentially microbial. Generally, hydrolysis is the major pathway in nonflooded soil while methyl parathion is degraded mainly by nitro-group reduction in predominantly anaerobic systems, such as flooded soil (Ou et al., 1983; Ou, 1985; Adhya et al., 1987).

In a few instances, hydrolysis is the major or only pathway of methyl parathion degradation in soils, even under flooded conditions (Ou, 1985). Adhya et al. (1987) evaluated the influence of different physical and chemical characteristics on the persistence of methyl parathion in 5 tropical soils under flooded and nonflooded conditions. They found that nitro-group reduction was the major pathway of methyl parathion degradation in 4 out of 5 of the soils under flooded conditions, while, in one soil (Sukinda-soil), degradation of methyl parathion proceeded exclusively by hydrolysis, even under flooded conditions. The latter finding was confirmed by Sharmila et al. (1989a).

A temperature-dependent shift from nitro-group reduction (at 25 °C) to predominantly hydrolysis (at 35 °C) occurred in a flooded alluvial soil; both pathways were mediated microbially (Sharmila et al., 1988). The addition of yeast extract also influenced the degradation pathway of methyl parathion by bacterial cultures in enriched flooded alluvial and laterite (Sukinda) soils (Sharmila et al., 1989b). Low redox potential in a flooded soil favoured degradation by nitro-group reduction, whereas hydrolysis was concomitant with a more positive potential (Adhya et al., 1981a).

Adhya et al. (1981b) reported studies on sulfur-containing anaerobic ecosystems, such as oceanic sediments, which they supposed could serve as a potential sink for pesticides. They found that methyl parathion was decomposed in acid, sulfur-containing soils and soils with a low sulfate content to aminomethyl parathion; however, no decomposition occurred under aerobic conditions. Demethylation could be demonstrated in anaerobic sulfate soils.

Evidence for microbial participation was provided by the fact that sterilization of the enriched soil samples increased the stability of methyl parathion in soil (Adhya et al., 1981a). The authors reported a very rapid reduction of the nitro group of methyl parathion by equilibration with a soil incubated with rice straw under flooding. Sterilization of this soil preparation prevented this rapid reduction. The degradation of methyl parathion and its metabolite *p*-nitrophenol in flooded alluvial soil is given in Table 12.

It appeared from this study, that the degradation of the metabolite *p*-nitrophenol is more rapid than the decomposition of methyl parathion.

Table 12. Degradation of methyl parathion and its metabolite *p*-nitrophenol in flooded alluvial soil[a]

Days after methyl parathion addition	μg of compound recovered/20 g of soil	
	methyl parathion	*p*-nitrophenol
0	485.3	0
0.5	428.1	trace
1	333.7	120.0
2	219.8	98.6
3	185.6	72.0
6	95.5	0
12	58.2	0

[a] From: Adhya et al. (1981a).

Isolated mixed bacterial cultures from soil utilized methyl parathion and parathion as a sole carbon source (Chaudhry et al., 1988). *Pseudomonas* sp. was capable of hydrolysing methyl parathion and parathion to *p*-nitrophenol but needed another carbon source for growth. The optimum pH range for enzymatic hydrolysis by this bacterium was from 7.5 to 9.5. In view of the instability of methyl parathion in alkaline solutions, it is not clear whether the hydrolysis noted was or was not partially due to the pH of the solution rather than wholly due to bacterial action. The thermal optimum was between 35 °C and 40 °C. *Flavobacterium* sp. culture was able to metabolize *p*-nitrophenol by degrading it to nitrite and to use it for growth. The DNAs from *Pseudomonas* sp. and from the mixed culture showed homology with the organophosphate degradation gene from a previously reported parathion-hydrolysing bacterium, *Flavobacterium* sp. Ou & Sharma (1989) showed that methyl parathion is extensively degraded by a mixed bacterial culture and a *Bacillus* sp. to its final oxidation products carbon dioxide and water, whilst a *Pseudomonas* sp. isolated from the mixed culture could degrade the hydrolysis product *p*-nitrophenol. A *Flavobacterium* sp. isolated from flooded soil was able to hydrolyse methyl parathion, but a *Pseudomonas* sp. from flooded soil was not (Adhya et al., 1981c). The transformation of methyl parathion by pure cultures of *Flavobacterium* sp. followed multiphasic kinetics (Lewis et al., 1985).

A different result was described by Arndt et al. (1981) for microorganisms in compost. They added 70 mg of methyl parathion dissolved in 20 ml ethyl acetate to 1.2 kg of grass (40%), apples (23%), potatoes (17%), yoghourt (13%), and bread (7%). After composting this mixture for 7 days, no degradation product of methyl parathion was found. The recovery rate was 95%. The authors concluded that the insecticide could accumulate in the compost under the conditions tested, but it could not be excluded that this result was affected by the ethyl acetate.

The concentration of methyl parathion (applied at 0.28 kg/ha) in a lake (Clear Lake, California, USA) dropped from 0.50 μg/litre to 0.28 μg/litre, measured 8 and 48 h, respectively, after treatment (Apperson et al., 1976). After a third application (total 3 × 0.28 kg/ha) the residue level of methyl parathion was 5.4 μg/litre, and 7 days later, 2 μg/litre (Apperson et al., 1976). Eichelberger & Lichtenberg (1971) found that 90% of methyl parathion in river water was degraded during a period of 2 weeks, whereas there was no degradation in distilled water. The latter finding may be pH related, since Cowart et al. (1971) noted 50% hydrolysis of the pesticide after 14 days in distilled water at pH 6. Under field conditions, in the presence of sediment and aquatic plants, degradation is accelerated and persistence is lower. Dortland (1980) showed that persistence decreased by a factor of 2-3 when sediment and plants were added to the aquatic microcosm. When considering the aquatic ecosystem as a whole (which includes adsorption on sediments and adsorption on, and incorporation in, aquatic biota) a fair estimate of the persistence of methyl parathion in the water column seemed to be 2-3 days (Walker, 1978). This value was recorded in microcosm studies and field experiments in both freshwater and estuarine aquatic environments. Predicted half-life values in rivers, ponds, eutrophic lakes, and oligotrophic lakes were reported to be 0.6, 27.3, 28.3, and 151.6 h, respectively (Smith et al., 1978). Methyl parathion was degraded with a half-life of 28 h in sediment collected from a field site and with a half-life of 7 h in microbial mats derived from laboratory mesocosms (Newton et al., 1990). The half-lives of methyl parathion in the water and sediment of a carp pond were 5.7 days and 5.0 days, respectively (initial residues: 3.77 mg/litre in water and 0.52 mg/kg in soil) (Sabharwal & Belsare, 1986). It should be emphasized that the persistence values reported depend not only on the type of biotope but also on the abiotic conditions, i.e., temperature, pH, and salinity, as pointed out, for example, by Badawy & El-Dib (1984).

Holm et al. (1983) found in their model ecosystem that the sediment type had no observable effect on the degradation of methyl parathion and that it depended primarily on the communities of microorganisms. These communities and their ability to degrade methyl parathion did not change with different sediment types. The microbial degradation rate constants in an aquatic channel micro-cosmos ranged from 2.7×10^{-6}/s to 6.9×10^{-6}/s. This was significantly higher than the rate constants determined for abiotic degradation. Cripe et al. (1987) modified the river die-away test for determining the biodegradability of organic substances and tested the degradation products for their toxicity. Because of their sensitivity, mysids and daphnids were used for testing the toxicity of the degradation products. This test showed a rapid, sediment-mediated biodegradation of methyl parathion.

The biodegration rate of methyl parathion was compared in 3 types of test systems composed of sediment and water collected from various estuarine sites (Van Veld & Spain, 1983). Generally, methyl parathion degradation was fastest in intact sediment/water cores, followed by sediment/water shake flasks, and was slowest in water shake flasks.

Lewis & Holm (1981) determined the transformation rate of methyl parathion by "aufwuchs" microorganisms, i.e., aquatic microbial growth attached to submerged surfaces or suspended in streamers or mats. "Aufwuchs" fungi, protozoa, and algae did not transform methyl parathion, but bacteria rapidly transformed it.

Lewis et al. (1984) examined the effects of microbial community interactions on methyl parathion transformation rates. They found either stimulation or inhibition of bacterial transformation rates in the presence of various cultures, filtrates, or exudates of algae, fungi, or other bacteria.

The biotic and abiotic degradation rates of methyl parathion in water and sediment samples over a 3-year period was studied by Pritchard et al. (1987). The aim of their study was to find the reason for the different degradation rates reported for methyl parathion, but the divergences in biodegration could not be assigned to any single factor. The predominant degradation in an aerobic system appears to be the biological hydrolysis, producing *p*-nitrophenol.

Phosphatases are an important group of enzymes involved in the breakdown of methyl parathion (Portier & Meyers, 1982; Portier et al., 1983). A proposed pathway for the breakdown of methyl parathion in aquatic systems is given by Bourquin et al. (1979) in Fig. 1.

Methyl parathion is degraded by bacteria in soil, but more slowly by bacteria in water. Crossland et al. (1986) estimated the rate of biodegradation of methyl parathion using a mathematical model. Sorption on sediment was the dominant process for loss of methyl parathion from the water compartment. The rate of biodegradation in sediment (4.0 μmol/litre per h) greatly exceeded that of sorption on sediment (0.02-0.05 μmol/litre per h) and, therefore, the sediment compartment may be considered a sink for methyl parathion.

The complete decomposition of methyl parathion into innocuous compounds can be realized by planktonic and attached microorganisms (Lassiter et al., 1986). The metabolite p-nitrophenol can be further metabolized by algae, as reported by Werner & Pawlitz (1978).

4.2.2 Abiotic degradation

Data on the abiotic degradation of methyl parathion are presented in Table 13.

4.2.2.1 Photodegradation

When exposed to UV radiation or sunlight, methyl parathion undergoes oxidative degradation. The degradation rate constant of methyl parathion sprayed as a film (0.67 μg/cm$_2$) and exposed to 300 nm light was reported to be 46.6×10^{-7}/s, corresponding to a half-life of 41.2 h (Chen et al., 1984). In a stationary reactor, the half-life of methyl parathion dissolved in an aqueous solution (pH = 7) was 72 min after radiation with a Hg low pressure lamp (at 254 nm) (Hicke & Thiemann, 1987). Methyl parathion has been shown to be one of the most light-sensitive insecticides. Baker & Applegate (1970, 1974) showed photodegradation of methyl parathion using light in the spectral range 300-400 nm (Table 13); methyl paraoxon, the active cholinesterase inhibitor, was produced. Although

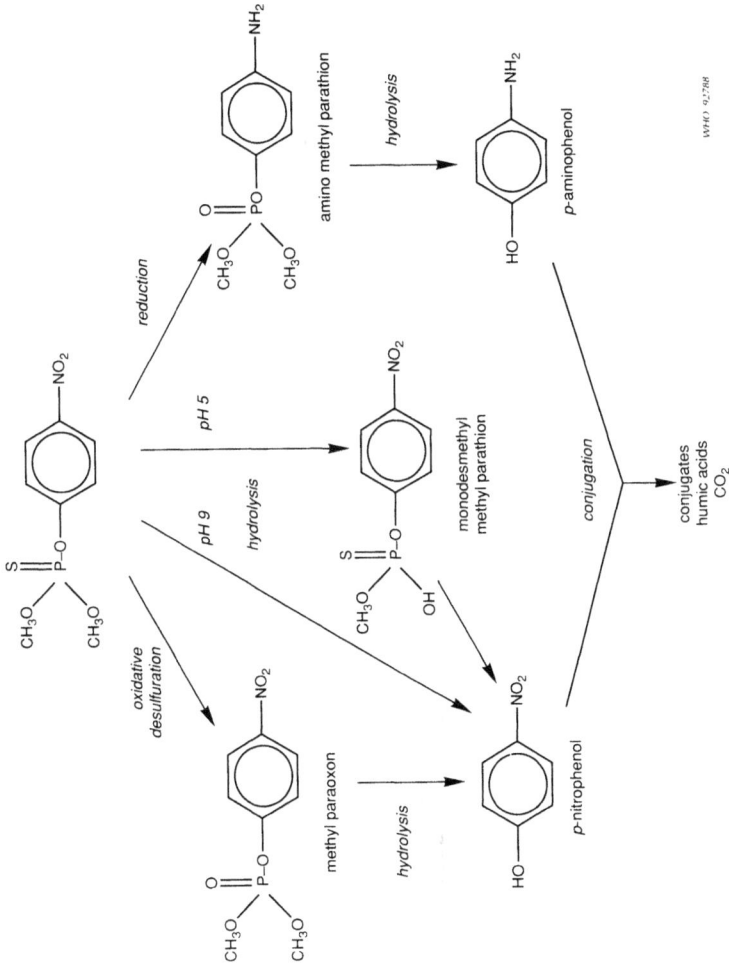

Fig. 1. Proposed pathway for the breakdown of methyl parathion in aquatic systems (modified from: Bourquin et al., 1979; Wilmes, 1987).

Table 13. Abiotic degradation of methyl parathion

Transformation process	Time	Experimental conditions temp (°C)	Experimental conditions pH	Light	Initial concentration (mg/litre)	Conversion (%)	References
Hydrolysis in distilled water	24 h	a	6		0.26	8.8	Cowart et al. (1971)
	7 days	a	6		0.26	32.0	Cowart et al. (1971)
	14 days	a	6		0.26	50.5	Cowart et al. (1971)
	21 days	a	6		0.26	73.8	Cowart et al. (1971)
	28 days	a	6		0.26	100	Cowart et al. (1971)
Hydrolysis in distilled water	31.7 days	10	1-5		a	50	Mühlmann & Schrader (1957)
	12.5 h	40	1-5		a	50	Mühlmann & Schrader (1957)
Hydrolysis in ethanol buffer	8.4 h	70	6		6	50	Ruzicka et al. (1967)
Hydrolysis in 0.01 M NaOH	4 h	37.5	12		a	64-73	Jaglan & Gunther (1970)

74

Table 13 (continued)

UV-degradation of pure product	2 h	30	350 nm	0.1	39	Baker & Applegate (1974)
	4 h	30	350 nm	0.1	65	Baker & Applegate (1974)
	6 h	30	350 nm	0.1	82	Baker & Applegate (1974)
	8 h	30	350 nm	0.1	91	Baker & Applegate (1974)
Temperature-degradation of pure product	2 h	35	dark	0.1	9	Baker & Applegate (1974)
	4 h	35	dark	0.1	8	Baker & Applegate (1974)
	6 h	35	dark	0.1	24	Baker & Applegate (1974)
	8 h	35	dark	0.1	31	Baker & Applegate (1974)

a No data given.

photodegradation of methyl parathion in the terrestrial compartment of the environment may be important, it plays only a minor role in aquatic media (Env. Res. Lab., 1981). The first-order transformation rate for photolysis upon exposure to daylight fluorescent lamps was low compared to hydrolysis and, in particular, compared to microbial degradation in an aquatic channel microcosm (Holm et al., 1983). The loss of methyl parathion through photolysis was estimated to be 4%.

Nevertheless, it seems that sunlight may reduce the half-life of methyl parathion considerably. Schimmel et al. (1983) reported a half-life of 6.3 days for a 1 mg methyl parathion/litre solution exposed to sunlight. In darkness, with the same test conditions, the half-life was 18 days. Like parathion, the photoreaction of methyl parathion was accelerated in the presence of green and blue green algae (Zepp & Schlotzhauer, 1983).

Exposure of methyl parathion to sunlight resulted in the formation of trace levels of *O, O, S*-trimethyl phosphorothioate and trimethylphosphate (Chukwudebe et al., 1989).

According to Sauvegrain (1980), methyl parathion seems to be oxidized by oxidizing agents, i.e., ozone and chlorine. Methyl parathion treatment with ozone eliminated 80-100% of the compound. The oxidation of methyl parathion leads to methyl paraoxon, which is further transformed into *p*-nitrophenol.

4.2.2.2 Hydrolytic degradation

The half-life of methyl parathion in an aqueous solution (20 °C, pH 1-5) was reported to be 175 days (Melnikov, 1971). At a concentration of 0.03 mol/litre (pH 10), sodium perborate greatly accelerated the degradation of methyl parathion (Qian et al., 1985). The half-life in the presence of perborate was 12 min, while the rate was too slow to be measurable when the same concentration of sodium carbonate was added. Badawy & El-Dib (1984) also found that the degradation of methyl parathion occurred much more rapidly under alkaline (pH 8.5) than under neutral (pH 7.0) or acidic (pH 0.5) conditions. The rate of degradation was also positively correlated with salinity.

Although chemical hydrolysis occurs in the aquatic environment, this degradation reaction plays only a limited role in the disappearance of methyl parathion. In an aquatic channel microcosm, only 7% of degradation of the pesticide was attributed to chemical hydrolysis (Holm et al., 1983). In a sterile, seawater-sediment system, methyl parathion remained for 7 days whereas, in a corresponding nonsterile system, 100% of the compound was degraded within this period (Env. Res. Lab., 1981).

Methyl paraoxon, the more toxic oxygen analogue of methyl parathion is also chemically hydrolysed. According to Jaglan & Gunther (1970), the chemical hydrolysis of methyl paraoxon is much faster than that of methyl parathion, because of the presence of oxygen in the oxon, which makes the phosphorus more susceptible to attack by the hydroxide ion. At pH 8.5 (37.5 °C), approximately 35% of methyl paraoxon was hydrolysed within 16 h compared with about 5% for methyl parathion.

The hydrolysis products of methyl parathion and methyl paraoxon are dimethyl phosphorothioic acid or dimethyl phosphoric acid and p-nitrophenol. These compounds are less toxic than the parent compounds, thus hydrolysis is detoxifying (Thuma et al., 1983).

Pritchard et al. (1987) reported that there was no biotic degradation of methyl parathion in seawater, i.e., "the rate resulting from the substraction of the sterile rate from the nonsterile rate was not significantly different from zero".

Several research groups investigated the binding of methyl parathion on soils as well as the soil catalysed degradation of methyl parathion. Saltzman et al. (1976) and Mingelgrin et al. (1977) analysed the influences of different water contents and cations on the kaolinite-catalysed degradation of methyl parathion; when adsorbed on kaolinite, methyl parathion seems to be more stable than parathion. A concentration of 10% Ca-kaolinite catalysed the degradation of methyl parathion most efficiently.

Wolfe et al. (1986) studied the influences of pH and redox transformations on the detoxification of methyl parathion in soils, quantitatively. The disappearence of methyl parathion could be described by first-order kinetics. Amino methyl parathion was identified as a reaction product. Half-lives in the range of a few minutes were measured in strongly reducing sediments, thus, confirming the data of Gambrell et al. (1984). It was suggested that more information about the effect of sediment sorption was needed for further studies on the reaction kinetics.

4.2.3 Bioaccumulation

Temporary accumulation (up to 10 days) occurred following an aerial spraying of pine and deciduous forest with methyl parathion (3 kg 20% solution/ha in April and 1 kg 40% solution/ha in September), which led to higher levels of methyl parathion in the tissues of a variety of vertebrates compared with the concentrations in soil, water, and plants (Fedorenko et al., 1981).

Takimoto. et al. (1984) reported bioaccumulation of methyl parathion in killifish (*Oryzias latipes*). Bioaccumulation factors of 88-fold (postlarva) to 540-fold (female adult) were found in the killifish. Residues in the bluegill sunfish (*Lepomis macrochirus*), exposed to methyl parathion treatments in a lake, varied from 11 to 110 μg/kg, corresponding to bioaccumulation factors of 28-39 (Apperson et al., 1976).

Sabharwal & Belsare (1986) added 4 mg methyl parathion/litre to the water of a carp-rearing pond and measured the methyl parathion concentrations in the water, soil, macrophytes, and carps over a period of 35 days. The methyl parathion limits of detection in water, soil, macrophytes, and fish were 0.0066, 0.12, 0.0478, and 0.0746 mg/kg respectively (see Table 14).

There was an accumulation of methyl parathion in the soil, macrophytes, and fish, whereas the compound degraded immediately in water. The bioaccumulation in the carp peaked at 3 days.

Using a mean K_{ow} value of 2.55, and on the basis of the log K_{ow}/log bio-concentration regression curve for fathead minnows, the estimated bioconcentration factor was reported to be 22 (Env. Res. Lab., 1981). According to Zitko & McLeese (1980), the expected bioconcentration factor in aquatic biota for methyl parathion is estimated to be 20.

Crossland & Bennett (1984) using a range of published log K_{ow} values estimated that bioaccumulation factors would be between 2.5 and 84.

Table 14.   The persistence of methyl parathion in water, soil, macrophytes, and fish[a]

Time (days)	methyl parathion concentration (mg/kg)			
	water	soil	macrophytes	fish
0	3.77	0.52	1.2	0.52
1	3.15	2.28	14.41	10.26
3	2.16	1.5	11.73	26.17
7	1.50	-	8.98	11.74
14	0.60	-	4.16	5.67
21	0.28	-	2.24	2.06
28	nd[b]	nd[b]	1.42	0.83
35	nd[b]	nd[b]	0.73	0.48

[a]From: Sabharwal & Belsare (1986).
[b]nd = not detectable.

Accumulation of methyl parathion does not occur in the blood of mammals. After ingestion, it is rapidly absorbed and the blood concentration reaches a maximum 1-3 h following ingestion and, thereafter, decreases. Although a significant portion of methyl parathion is found in the bile, it is present in all organs (see also section 6.2).

4.3 Interaction with other physical, chemical, and biological factors

Methyl parathion shows interactions with the following substances: adrenocorticoids, anaesthetics, tricyclic antidepressive agents, antihistamines, atropine, barbiturates, clofibrate, colistimethate, corticosteroids, curare, decamethonium, dexpanthenol, fluorophosphate, hexamethonium, kanamycin, morphine, muscle relaxants, anticholinesterases, neomycin, parasympathomimetics, phenothiazines, polymyxin, pralidoxime, procainamide, streptomycin, succinylcholin, sympathomimetics, d-tubocurarine (Martin, 1978).

A significant increase in the toxicity of oxygen analogues of organophosphorus insecticides to house flies was observed following treatment with polychlorinated biphenyl (PCB) (Aroclor 1248)

(Fuhremann, 1980).   Detergents increased the hydrolysis of organophosphates, such as methyl parathion (Peterka & Cerna, 1988).   Yang & Sun (1977) found an inversely proportional correlation between fish toxicity and the partition coefficient of different insecticides, including methyl parathion.

DEF (*S,S,S*-tributyltrithiophosphate), a defoliant, enhanced the toxic effect of methyl parathion in the fish (*Gambusia affinis*) (Fabacher, 1976).

4.4 Ultimate fate following use

The ultimate fate of methyl parathion depends on the degradation pathways. The most important one is chemical as well as biological hydrolysis; the others are oxidative desulfurisation, nitro reduction, and photodegradation. Important degradation products are methyl paraoxon, dimethylthiophosphoric acid, dimethylphosphoric acid, and *p*-nitrophenol.

5. ENVIRONMENTAL LEVELS AND HUMAN EXPOSURE

5.1 Environmental levels

5.1.1 Air

In a pilot study, Stanley et al. (1971) measured methyl parathion concentrations of up to 129 ng/m^3 in air samples collected in the USA (Stoneville). The technique used for air sampling was that of Miles et al. (1970).

In Tennessee, USA, average hourly concentrations of methyl parathion in air were < 0.57 ng/m^3 (maximum, 2.9 ng/m^3) at a site located one mile south-east of a methyl parathion plant and one mile west of a plant producing the nematocide ethoprophos (*O*-ethyl-*S*,*S*-dipropyl phosphorodithioate), and < 0.64 ng/m^3 (maximum, 5.1 ng/m^3) at another site located one mile north of a methyl parathion plant. Particulate samples collected from the 2 sites contained < 0.086 ng methyl parathion/m^3 (Foster, 1974).

In the USA, maximum atmospheric levels were detected of 29.6 ng/m^3 in Alabama, 5.4 ng/m^3 in Florida, and 129 ng/m^3 in Mississippi (Midwest Research Institute, 1975). Methyl parathion was found in air samples in the Mississippi Delta, one of the highest pesticide usage areas in the USA, because of the intensive cotton production, at a maximal concentration of 2060 ng/m^3 (Arthur et al., 1976). The average monthly concentrations of methyl parathion peaked in August or September with levels varying from 111.7 ng/m^3 (September 1972) to 791.1 ng/m^3 (September 1973).

In another study, airborne residues of methyl parathion and methyl paraoxon were determined after the use of methyl parathion on rice in the Sacramento valley in California, USA (Seiber et al., 1989). Sampling was conducted on the roof tops of public buildings in 4 towns in 2 counties where methyl parathion was used in significant quantities, and in a reference area where no use occurred. Daily maximum average concentrations were 25.7 ng/m^3 for methyl parathion and 3.1 ng/m^3 for methyl paraoxon. The range in averages for all sites in the vicinity of usage during springtime 1986 was 0.2-6.2 ng/m^3 for methyl parathion and < 0.5-0.8 ng/m^3 for methyl paraoxon. With one exception, the background samples did not show any methyl parathion above the detection limit.

Methyl parathion and methyl paraoxon concentrations measured in the condensate from coastal fog near Monterey (California, USA) ranged between 0.046 and 0.43 μg/litre and between 0.039 and 0.49 μg/litre, respectively. The oxon to thion ratios were 0.28-2.6, and thion to oxon conversion appeared to take place during atmospheric transport from agricultural to the nonagricultural areas (Schomburg et al., 1991).

In the Kalinin District, Tashkent Province, the Uzbek SSR (USSR), during July and August, the concentrations of methyl parathion in the air after spraying with 30% emulsion, measured at 500, 750, and 1000 m from the place of the treatment, were 0.055-0.08, 0.01-0.02, and 0-0.008 mg/m^3, respectively (Akhmedov, 1968).

Tessari & Spencer (1971) analysed indoor and outdoor air samples, collected monthly for a year, at the homes of families where the head of the household was occupationally exposed to pesticides. A nylon chiffon cloth screen was exposed to the atmosphere for 5 days and the absorbed pesticides were extracted and analysed using a column chromatography method. The authors found methyl parathion in 13 out of 52 samples, at an average concentration of 1.04 μg/m^3. The range was 0.04-9.4 μg/m^3. The values obtained from outdoor sampling were much smaller, 3 out of 53 samples containing 0.35 μg methyl parathion/m^3 with a range of 0.15-0.71 μg/m^3.

5.1.2 Water

Methyl parathion concentrations of up to 0.23 μg/litre were found in selected Western streams of the USA in 1968-71 (Schulze et al., 1973).

In 1970, methyl parathion was detected in 3 out of 18 surface drain effluent water samples in California, USA, at concentrations of 10-190 ng/kg, and, in 8 out of 60 subsurface drain effluent water samples, at concentrations of 10-170 ng/kg (Midwest Research Institute, 1975).

In water samples from 10 sites in the Cape Fear River Basin in North Carolina, USA, taken monthly between July 1974 and June 1975 (except October), maximum concentrations of methyl parathion in dissolved fractions and in particulate-associated fractions were

468 ng/litre and 123 ng/litre, respectively (Pfaender et al., 1977). Methyl parathion was detected in waste water from a parathion production plant in the USA at levels of 2.0 mg/litre in pre-treatment water and < 0.004 mg/litre in post-treatment water (Marcus et al., 1978).

Methyl parathion residues in major Mississippi stream systems (USA), monitored during 1972-73, ranged between 0.08 and 0.46 μg/litre (Leard et al., 1980).

In one station at the Negro River Basin (Argentina), methyl parathion was detected at a concentration of 0.034 μg/litre in March 1986, which is the end of the summer season in South America (Natale et al., 1988).

In a study on the Ionnina basin and Kalamas river (Greece), from September 1984 to October 1985, a seasonal fluctuation was found in the concentration of methyl parathion, with a maximum during the summer and a minimum during the winter (Albanis et al., 1986). The mean concentration in the lake Pamvotis (Greece) was 7.7 ng methyl parathion/litre in July. The natural outlet of the lake is the Kalamas River, where a maximum concentration of 32 ng methyl parathion/litre was found. With the exception of the river, the other analyses showed much lower concentrations of methyl parathion. The results of this study show very clearly the seasonal influence of the application of this pesticide on natural water concentrations.

Normally, the methyl parathion concentration in the River Rhine is below the limit of detection and the Sandoz accident on 1 November 1987 did not affect the wells of the waterworks. A maximum value measured in the Rhine during the second half of 1986 was higher (< 0.05 mg/m^3) than that following this accident (Winter & Lindner, 1987).

Methyl parathion was detected in Hungarian surface waters only once between 1977 and 1986 (concentration not given), which corresponded to a sampling frequency of 0.14% (Csernatoni et al., 1988).

5.1.3 Soil

In 1969, 76 samples of onions and the soils in which they had been grown were collected in the 10 major onion-producing states of the USA for analysis of the pesticide residues. The limit of quantification of methyl parathion was 0.01 mg/kg. Methyl parathion was found in a range of 0.09-1.9 mg/kg in 11.8% of the soil samples. No residues were detected in the onion samples (Wiersma et al., 1972).

Methyl parathion was found at levels of 0.09-1.90 mg/kg in soil samples from onion-producing States in the USA (Midwest Research Institute, 1975). In cropland soil (South Dakota, USA), the concentration of methyl parathion was 0.01 mg/kg soil (Carey et al., 1979).

5.1.4 Food

Renvall et al. (1975) reported pesticide analyses of fruits and vegetables on the Swedish market from July 1967 to April 1973. Methyl parathion belonged to the most frequently occurring pesticides with a rate of 6%. Levels in 4 out of 207 oranges analysed, 1 out of 37 lemons, 4 out of 69 grapefruits, and, 2 out of 29 clementines or mandarins exceeded 0.11 mg/kg. In a more recent study, in the Swedish monitoring programme during the period 1981-84, methyl parathion was found in apples, celery, grapes, lemons, lettuce, limes, mandarins, oranges, pears, and plums. One out of 74 celeries analysed (imported), 1 out of 238 lemons (imported), 1 out of 248 lettuces (domestic), 5 out of 421 mandarins (imported), and, 8 out of 917 oranges (imported) exceeded the Swedish maximum residue limits of 0.1-0.5 mg methyl parathion/kg (Andersson, 1986).

In a study on the presence of organophosphorus insecticide residues in Mexican food, methyl parathion residues were found in market samples of avocados, rice, strawberries, and tomatoes, with respectively 6, 4, 3, and 5 positive samples out of 10. The average concentrations were 0.3, 0.8, 0.5, and 0.5 mg/kg, respectively (Albert et al., 1979).

A report on pesticide residues in the United Kingdom (1982-85) gave a residue level for methyl parathion in lemons of 0.3 mg/kg (MAFF, 1986). In a more recent report, no methyl parathion was found in cooking apples and in imported apples with a reporting limit of determination of 0.1 mg/kg; however, a concentration of 0.08 mg methyl parathion/kg was found in one sample of lemons from Spain (MAFF, 1990).

Methyl parathion was detected in citrus fruits in France at levels of 0.003-1.25 mg/kg (Mestres et al., 1977). Lamontagne (1978) found methyl parathion in concentrations of 0.311 mg/kg in fruit and 0.87-2.12 mg/kg in greenhouse plants in France. Branca & Quaglino (1988) found methyl parathion at a residue level of 0.036 mg/kg in one out of 34 samples of French potatoes imported into Italy.

Pesticide residue levels were analysed during 1968-69 in samples of ready-to-eat foods from 30 markets in 24 different cities with populations of between 50 000 and more than 1 000 000 in the USA. The limit of determination was 0.05 mg/kg. Methyl parathion was found infrequently (1 × Boston, 1 × Los Angeles, 2 × Minneapolis) in concentrations of 0.008, traces, 0.001, and 0.025 mg/kg in leafy vegetables and 0.033 mg/kg in grain (Boston) (Corneliussen, 1970). From June 1971 to July 1972, methyl parathion was detected in 7 out of 420 samples of ready-to-eat foods. The concentrations found in leafy vegetables ranged from a trace to 0.010 mg/kg. In one sample of fruit (type not given), a concentration of 0.007 mg/kg was found (Boston) (Manske & Johnson, 1975). In the report of the Food and Drug Administration, 5 samples of leafy vegetables containing methyl parathion residues are mentioned. The concentrations ranged from a trace to 0.003 mg/kg (Johnson & Manske, 1976). In "market-basket" surveys conducted by the US Food and Drug Administration in 1966-69, methyl parathion was detected in leafy and stem vegetables at levels of 0-2.00 mg/kg, and, in root vegetables, at levels of 0-1.0 mg/kg (Midwest Research Institute, 1975). Johnson et al. (1981) did not find any methyl parathion in infant and toddler Total Diet Studies (TDS) in the USA in 1975-76. In the adult TDS in the USA in 1973-74, trace residue levels were found in leafy vegetables, but none in fruit (Manske & Johnson, 1977). "Dislodgable" methyl parathion residues were found on sweet corn in the USA at levels of 0-0.14 $\mu g/cm^2$, one and two days after application of the pesticide (Wicker et al., 1979). Soybeans analysed in 1979 showed levels of 1-40 mg methyl parathion/kg and soybean forage analysed at intervals of 1-14 days after treatment, 0.3-6.6 mg methyl parathion/kg. Levels of 0.1-0.3 mg methyl parathion/kg were measured in 12 samples of cottonseed (FAO, 1985).

Samples of standing agricultural crops were analysed in 1971 during the National Pesticide Monitoring Programme in the USA (Carey et al., 1978). Levels of methyl parathion detected in samples of alfalfa, field orn (kernels), cotton, cotton stalks, and mixed hay ranged from 0.02 to 4.57 mg/kg dry weight.

During a TDS in Canada in 1972, Smith et al. (1975) found methyl parathion residues in leafy vegetables from Winnipeg at an average level of 0.012 mg/kg.

In a TDS in New Zealand during 1971-73, methyl parathion was found in one sample of leafy vegetables at a level of 0.15 mg/kg in 1973, in one sample of root vegetables at the level of 0.26 mg/kg in 1972, and in 4 samples of citrus fruit at an average level of 0.20 mg/kg and a maximum level of 1.4 mg/kg during each of the years 1971-73.. In 1971, 3 samples of pip fruit contained, on average, 0.03 mg/kg, and, in 1972, one sample of stone fruit contained 0.25 mg/kg. Some of these figures exceeded the New Zealand tolerances (Love et al., 1974). In 1974, methyl parathion was detected at levels of 0.003-0.007 mg/kg in fruit and 0.002-0.008 mg/kg in tinned food from Auckland and Wellington, New Zealand (Dick et al., 1978).

The loss of methyl parathion in food during heating and storage was confirmed by Elkins et al. (1972). The samples were analysed before, and after, standardized heat treatment. Spinach and apricots were fortified separately with methyl parathion. The spinach samples were heated for 66 min at 122 °C and the apricot samples were heated for 50 min at 103 °C. The initial concentration of methyl parathion in the spinach samples was 0.88 mg/kg. It disappeared completely after heating. The methyl parathion level in the apricot samples was 0.85 mg/kg, but this decreased to 46% of this level after heating. The detection limit was less than 0.005 mg/kg. A further decomposition can be expected during the storage of preserved food. Generally, methyl parathion residues in fruit decomposed very rapidly, except in the waxy skin of apples and in the oil vessels of olives (Stoll, 1982).

Rippel et al. (1970) found remarkable differences in the degradation of methyl parathion in packaged citrus juice, depending on the kind of package surface. The rate of decrease of the methyl parathion residues was insignificant in glass containers. It was substantially higher in packages with tin-layer surfaces than in packages with painted protective surfaces, since the tin layers reduced the nitro group of the methyl parathion.

5.1.5 *Terrestrial and aquatic organisms*

Methyl parathion is rapidly metabolized in most organisms, resulting in low bioconcentration factors after acute exposure. There are few studies of residues of methyl parathion in organisms in the environment, but those conducted have consistently shown low methyl parathion residues.

Methyl parathion was detected in tissue samples from estuarine fish at a mean level of 47 μg/kg (Butler & Schutzmann, 1978). It has been detected at a concentration of 59 μg/kg in the ovaries of spotted sea trout (*Cynoscion nebulosus*), collected in Texas, USA (Midwest Research Institute, 1975).

Methyl parathion was detected in 34 out of 55 suspectedly poisoned apiaries examined in Connecticut (USA) in 1983-85 (Anderson & Wojtas, 1986). Concentrations of methyl parathion found in dead bees and in brood comb ranged from 0.04 to 5.8 mg/kg.

5.2 General population exposure

The general population can come into contact with methyl parathion via air, water, or food. Average methyl parathion intake from food in the USA during 1988 was estimated to range from 0.1 to 0.2 ng/kg per day in 3 different age groups (FDA, 1989). Draper & Street (1981) estimated that a 70-kg male living in a residence adjacent (50 yards) to an alfalfa field sprayed with methyl parathion at a rate of 0.19 kg a.i./ha would be exposed to a total dermal dose of 0.38 mg. Within a pesticide monitoring programme in the USA, based on the analysis of 6990 samples collected from the general population via the National Center for Health Statistics 1976-80, *para*-nitrophenol as an indicator for exposure to methyl and ethyl parathion was detected in 2.4% of urine samples from 12 to 74-year-old persons (Carey & Kutz, 1985).

5.3 Occupational exposure during manufacture, formulation, or use

There is a special risk for farm workers, since incidents of poisonings and illnesses during the mixing, loading, and application of methyl parathion have been reported. Exposure may also occur during the cleaning and repair of equipment and during early re-entry into fields. According to NIOSH (1976), 150 000 workers in the USA (field workers, aerial application personnel, mixer and blender operators, tractor tank loaders, ground applicator vehicle drivers, field inspectors, and warehouse personnel) are conceivably exposed to methyl parathion. A maximum air concentration of methyl parathion was estimated to be 1.77 $\mu g/m^3$. The exposure to methyl parathion was estimated by Hayes (1971) for workers checking cotton for insect damage as 0.7 mg/h via skin contact and < 0.01 mg/h through inhalation (NIOSH, 1976).

Davis et al. (1981) estimated that workers in apple orchards sprayed with methyl parathion would be exposed to dermal doses ranging from 0.055 mg to 3.1 mg, with the amount varying with time after spraying and the formulation of the pesticide. Two field studies were carried out by Kummer & Van Sittert (1986) to evaluate the health risk for the farm workers. In a number of cases, the men involved in hand-held ULV-spraying wore very little clothing and did not stop spraying, when it was too windy. Another possible contamination risk was the filling of bottles from larger (25-litre) containers, and the repairing and cleaning of the equipment with unprotected hands. However, no signs of acute poisoning could be observed in any of the persons involved in these studies. The urine was collected in spot samples in one of the studies and in 24-h samples in the other. Methyl parathion absorption could be verified from its metabolites in the spraymen's urine. Average levels of urinary nitrophenol (mg/g creatinine) for 6 supervisors and 2 groups of sprayers were reported to be 0.08 (range of 0.05-0.20), 0.38 (range of 0.04-1.38), and 0.13 (range of 0.06-0.44), respectively. An intake of 0.4-13 mg methyl parathion was calculated from the excreted p-nitrophenol.

Since investigations showed that clothing worn by agricultural workers became contaminated with methyl parathion following application and that the laundering of contaminated clothing with uncontaminated fabrics resulted in the transfer of the methyl parathion residue, recommendations were made that contaminated fabrics should not be washed with regular family laundry. Suggestions for the procedure of laundering were made by Easley et al. (1981) and Laughlin et al. (1981). The most effective procedure was using a pre-rinse programme and a detergent together with sodium hypochlorite (NaOCl) as a bleach. Laughlin & Gold (1989) discussed further aspects of laundering protective clothing contaminated with methyl parathion. Fluorocarbon soil repellent finishes on such protective clothing decrease pesticide absorption, but may hinder pesticide removal in laundering. Storage of laundered garments at 20 °C with air flow and/or at high humidity levels was recommended to dissipate residues of methyl parathion.

Ware et al. (1974b) suggested that serum insecticide levels, serum and red blood cell cholinesterase activities, and urinary excretion of *p*-nitrophenol should be investigated, because they are more effective for evaluating the possible potential poisoning hazard than the analysis of skin and clothing contamination. The safety of re-entering cotton fields 24 h following application of methyl parathion was tested. Methyl parathion was applied at 1.12 kg a.i./ha. During the application, the temperature ranged from 30 to 38 °C. The foliar residues decreased from 1.6 mg/m^2, 24 h following methyl parathion treatment, to 0.9 mg/m^2, 6 h later. No methyl parathion was detectable in the serum of the volunteers. The 48-h urinary excretion of *p*-nitrophenol ranged from 0.15 to 1.20 mg. Serum cholinesterase levels varied within normal intervals whereas the red blood cell cholinesterase levels showed a temporary, but not pronounced, depression of about 5-7%. The amounts of methyl parathion and methyl paraoxon extracted from clothing and hand surfaces are shown in Table 15.

Table 15.   Extracted residues of methyl parathion and methyl paraoxon following a 5-h working period[a]

Extract from:	Methyl parathion residue (mg)	Methyl paraoxon residue (mg)
Hands	0.2	0.5
Shirts	0.2	4.0
mep.5s	1.7	39.0

[a] From: Ware et al. (1974b).

During the working period, the mean air concentration was 0.2 ng methyl parathion/litre, of which, 1.2 µg methyl parathion was inhaled over 5 h. From all these data, it was concluded that a 24-h interval is safe for methyl parathion in this form of application.

Munn et al. (1985) collected human exposure samples from workers and dependants wearing nylon gloves, as well as environmental samples, during the onion harvest season of 1982 in Colorado, USA. Children in agricultural settings normally accompany their parents to the fields, as part of a family unit, the young children playing in this environment and older children helping their parents in the fields. Munn et al. (1985) recorded the length of time the gloves were worn, and the age and sex of the participants. No association between age and methyl parathion levels was found. The urine samples collected prior to their leaving the field did not contain detectable levels of methyl parathion. This could be because the nylon gloves reduced the absorption of organophosphate residues by about 90%.

6. KINETICS AND METABOLISM

6.1 Absorption

Methyl parathion can be absorbed through the digestive tract, the skin, and the respiratory tract (White-Stevens, 1971).

The primary routes of exposure are via skin contact with contaminated plants or material, and via inhalation. Severe accidental intoxications of humans have occurred.

The absorption of methyl parathion from the digestive tract is rapid, and it appears in the bloodstream immediately after oral intake. Studies on guinea-pigs were performed to analyse the rate of absorption of radioactive labelled (^{32}P) methyl parathion. One minute after dosage, it could be detected in various organs. The maximum level was found 1-2 h after treatment. The liver showed a remarkably high concentration (Gar et al., 1958).

Miyamoto et al. (1963) administered 50 mg ^{32}P-labelled methyl parathion/kg body weight to guinea-pigs or 1.5 mg/kg body weight to rats, by stomach tube. Maximum concentrations in the blood and brain were reached 1-3 h after treatment. An oral dose of 50 mg methyl parathion/kg resulted in no detectable levels of methyl parathion in either the brain or blood after 3 min, but, after 6-8 min, at which point lethal effects occurred, levels of methyl parathion increased to 182 ng/ml in plasma and to 137 ng/g in brain (Yamamoto et al., 1981).

6.2 Distribution

Accumulation of methyl parathion was observed in tissues. The highest concentrations were found in the lung and the liver (NRC, 1977). Transplacental transport of methyl parathion is discussed in section 8.5.

Total radioactive residues recovered in the 12 tissues analysed (excluding the gastrointestinal tract) from rats given a single oral dose of 5 mg C-^{14}-methyl parathion/kg body weight were about 11% of the administered dose, 1 h after treatment, declining to 0.3% at 24 h, about 0.1% at 48 h, and to only 0.04%, 6 days later. The kidney had the highest relative activity up to 8 h after treatment. The ^{14}C-activity in the plasma was initially about 5 times higher than that in the erythrocytes. However, from day 2 to day 6 after dosing, the ^{14}C-activity in the erythrocytes was greater than that in plasma and remained constant (Weber et al., 1979).

Sultatos et al. (1990) measured the partition coefficient for methyl parathion between mouse liver and blood by either equilibrium dialysis or a perfusion technique and obtained values of 9.5 and 16.4 respectively.

In a kinetic study on mongrel dogs of both sexes, Braeckman et al. (1980) found a rapid decrease in serum methyl parathion concentrations during the first few hours. The authors injected methyl parathion intravenously in doses of 1, 3, 10, and 30 mg/kg body weight. The dogs were pretreated with 1-5 mg atropine/kg body weight, 10 min before injecting methyl parathion. The blood samples were taken for up to 160 h. Besides quantifying serum levels of methyl parathion, the authors also measured serum cholinesterase activity at the 2 highest concentrations of methyl parathion. The determination of serum methyl parathion concentrations was performed according to De Potter et al. (1978). The cholinesterase activity decreased within 30 min to its lowest value, i.e., 40% of the normal level in dogs receiving 10 mg/kg body weight and 25% in dogs receiving 30 mg/kg. The first rapid fall in the methyl parathion concentration after injection was due to distribution and elimination. A slower decrease in serum methyl parathion concentrations at higher doses was the result of deep compartment linear kinetics. This is in line with observations of Tilstone et al. (1979), who found a rebound effect after a haemoperfusion.

6.3 Metabolic transformation

Organic nitro compounds, orally administered to ruminants, will undergo reduction of the nitro groups to amino groups. This reaction takes place in the rumen (Karlog et al., 1978).

The metabolism of methyl parathion in rodents is illustrated in Fig. 2.

Because of the importance of a first pass through the liver for the metabolism of methyl parathion, there is a distinct difference between the oral and intravenous toxicity (Morgan et al., 1977; Braeckman et al., 1983). Conversion of methyl parathion to its toxic metabolite, methyl paraoxon, may occur within minutes following oral administration (Yamamoto et al., 1983).

Mouse liver, perfused with methyl parathion, released the toxic metabolite methyl paraoxon into the effluate. Mouse whole blood rapidly detoxified the methyl paraoxon formed (Sultatos, 1987).

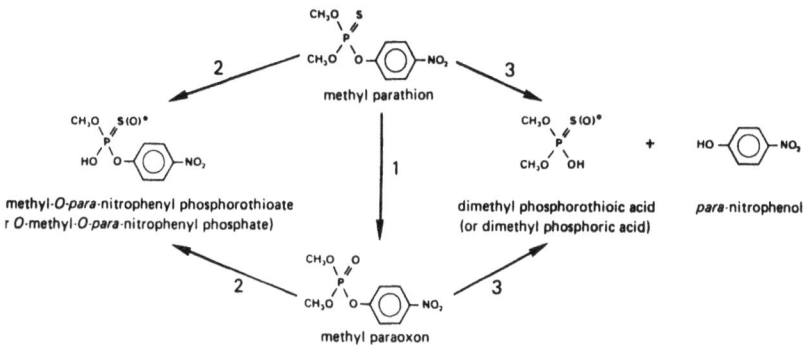

Fig. 2.   Metabolism of methyl parathion in rodents. Methyl paraoxon may be metabolized via the same pathways as methyl parathion, resulting in the oxygen analogue, indicated by the presence of (O)* in the figure. From: IARC (1983).

1. Toxification: metabolic formation of oxon.
2. Detoxification glutathione dependent alkyl transferase.
3. Detoxification glutathione dependent aryl transferase.

A reduction of the cellular concentration of reduced glutathione (GSH) influences mitosis, mobility, and other GSH-dependent cell functions. Glutathione S-transferases are mainly located in the cytosol and display overlapping substrate specificity. They also show peroxidase activity and prevent the peroxidation of membrane lipids. The interaction of methyl parathion with GSH or with the glutathione S-transferases therefore is important not only for the non-oxidative detoxification of the insecticide, but also for species-selective toxicity, and the development of resistance. Placental and fetal human glutathione S-transferase catalysed the dealkylation of methyl parathion exclusively to demethyl parathion via *O*-dealkylation (Radulovic et al., 1986; 1987).

Only after the metabolic formation of methyl paraoxon by liver microsomal oxidases does the substance become toxic. Therefore, this is an activation reaction. Methyl parathion and methyl paraoxon are mainly detoxified by conjugation with GSH (Hennighausen, 1984).

Detoxification is achieved by degradation reactions, that involve either demethylation or dearylation. The resulting desmethyl compounds and dimethyl phosphoric acids are essentially nontoxic (NRC, 1977). These detoxification reactions are due to the glutathione-dependent alkyl and aryl transferases; the reaction products are *O*-methyl-*O*-*p*-nitrophenyl phosphorothioate (or *O*-methyl-*O*-*p*-nitrophenyl phosphate) or dimethyl phosphorothioic acid (or dimethyl phosphoric acid) and *p*-nitrophenol. In addition, hydrolysis of methylparaoxon by tissue arylesterases may occur. Thus, it is possible to follow an exposure to methyl parathion by measuring the urinary excretion of *p*-nitrophenol (Benke & Murphy, 1975).

However, prior depletion of glutathione by acetaminophen (Costa & Murphy, 1984) or diethyl maleate (Sultatos & Woods, 1988) has little effect on the toxicity of methyl parathion in the mouse, indicating that perhaps glutathione does not play a significant role in the detoxification of methyl parathion.

The amount of the active toxic compound (methyl paraoxon) that will be produced after exposure to methyl parathion, depends on the kinetics of the oxidation of methyl parathion and on the kinetics of the detoxification reactions. Dealkylation is important at high dosages (Plapp & Casida, 1958). This enzyme system was found in the supernatant of the liver homogenate. The main metabolites were demethyl parathion (80%) and demethyl paraoxon (Fukami & Shihido, 1963; Shihido & Fukami, 1963).

The same major metabolites were generated when rat liver microsomes metabolized methyl parathion: demethyl paraoxon, methyl paraoxon, i.e., dimethyl phosphate, dimethyl phosphorothioate, and *p*-nitrophenol. When rats were treated with methyl parathion, dimethyl phosphoric acid was excreted in the urine together with *O*-methyl and *O,O*-dimethyl paraoxon (Menzie, 1974).

Adult rats have an increasing capacity to metabolize the oxygen analogue by both oxidative and hydrolytic pathways (Benke & Murphy, 1975).

Willems et al. (1980) calculated the high serum clearance of methyl parathion from their intravenous studies on dogs to be 2.1 litre/kg per h.

Malaysian prawns (*Macrobrachium rosenbergii*) as well as ridgeback prawns (*Sicyonia ingentis*) decomposed methyl parathion readily to *p*-nitrophenol and *p*-nitrophenyl conjugates. The dominant way of detoxification was the formation of β-glycosides and sulfate esters (Foster & Crosby, 1987).

The metabolism of methyl parathion in humans is similar to that reported in experimental animals (Fig. 3) (Benke & Murphy, 1975; Morgan et al., 1977). The liver is the primary organ for detoxification and metabolism (Nakatsugawa et al., 1968, 1969). The main metabolites recovered from urine following administration of methyl parathion to human subjects were also *p*-nitrophenol and dimethyl phosphate. Eight hours after application, *p*-nitrophenol excretion was nearly complete. Methyl paraoxon was hydrolysed to dimethyl phosphate and an amount representing 12% of the administered dose was excreted. Its excretion was more protracted than that of *p*-nitrophenol (Morgan et al., 1977).

Rao & McKinley (1969) found remarkable differences in the rates of metabolism of methyl parathion by liver homogenates from male and female chickens. The rate of the oxidative desulfurating system of the male liver homogenates was substantially higher than that of the homogenates of female chicken livers; however, the rates of the demethylating system showed no differences. Also no sexually determined differences of the oxidative or the demethylating system were found in the liver homogenates of rats, guinea-pigs, or monkeys.

6.4 Elimination and excretion in expired air, faeces, urine

After an oral dose of ^{32}P-methyl parathion to mice (17 mg/kg), 75% of the radioactivity was found after 72 h as metabolites in the urine and up to 10% was eliminated in the faeces (Hollingworth et al., 1967).

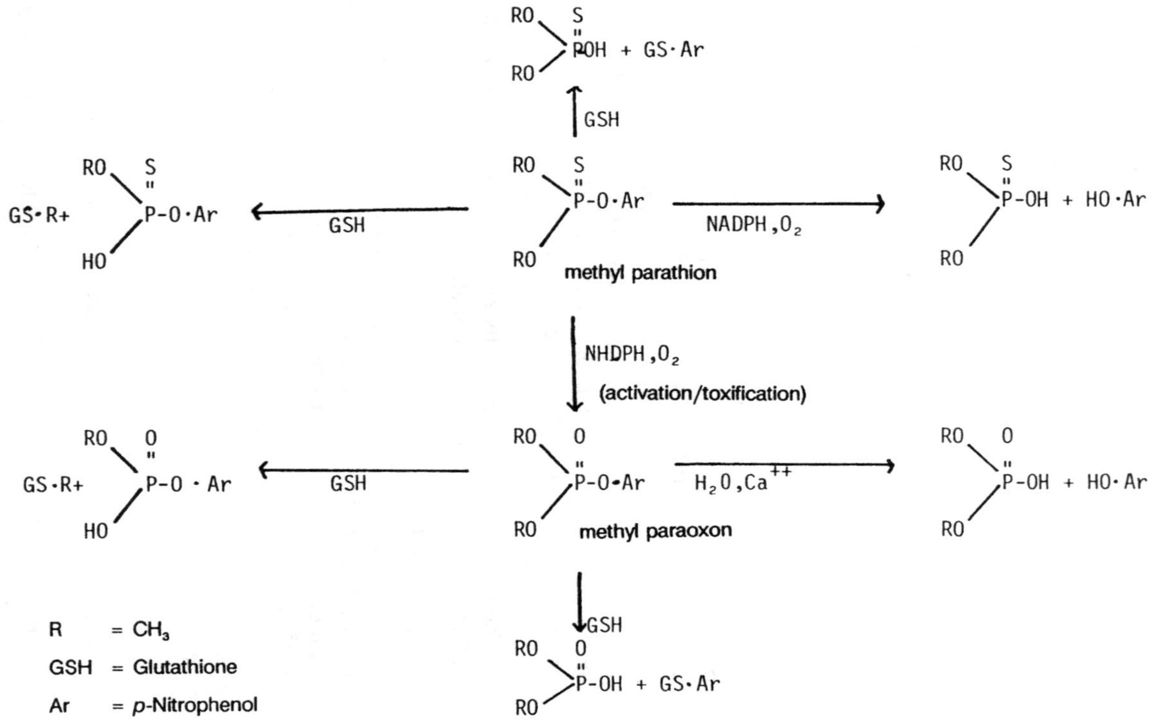

Fig. 3. Metabolism of methyl parathion in mammals. From: Flucke (1984).

In male rats, treated with a single oral dose of ^{14}C-methyl parathion (benzene ring-labelled) at 0.1, 1 or 5 mg/kg body weight and in female rats given a single oral dose of 1 mg/kg body weight, over 99% of the administered dose was eliminated in the urine and the faeces within 48 h. Elimination in the faeces accounted for only 5-7% after 1 or 5 mg/kg body weight, but amounted to about 20% after 0.1 mg/kg body weight Male rats treated with an intravenous dose of 1 mg methyl parathion/kg body weight eliminated about 99% of the administered radioactivity in the urine within 48 h, and approximately 1% of the dose in the faeces (Table 16, Weber et al., 1979).

Table 16. Elimination of ^{14}C-labelled methyl parathion in rats[a,b]

Doses (mg/kg)	Route of administration	No. of rats	Urine (%)	Faeces (%)	Balance (%)
0.1	oral	5	79.8 ± 11	19.4 ± 5.3	99.2
1	oral	4	93.6 ± 2.6	6.3 ± 1.1	99.9
1	iv	5	99.0 ± 3.8	0.8 ± 0.1	99.8
1	oral	4	93.3 ± 5.1	6.6 ± 2.7	99.9
5	oral	5	94.7 ± 6.0	5.1 ± 0.5	99.8

[a] Adapted from: Weber et al. (1979).
[b] iv = intravenously.

The kinetics of the toxic metabolite of methyl parathion, methyl paraoxon, were studied in conscious dogs (De Schryver et al., 1987). Thirty min before performing the test, the dogs received atropine as protection against intoxication. Methyl paraoxon was administered intravenously (2.5 mg/kg body weight) or orally (15 mg/kg body weight). The distribution of an intravenous dose was very fast.

The elimination was fitted by using a one-compartment model. The average half-life was determined to be 9.7 min, the average volume of distribution 1.76 litre/kg, and the average plasma clearance 126 ml/kg per min. Within 3-16 min, the maximal plasma concentration (927-2905 µg/litre) was reached following oral application. The bioavailability ranged from 5 to 71%. The hepatic extraction in anaesthetized dogs varied at a high level of 70-92%. From comparison of the urinary excretion as *p*-nitrophenol after intravenous (87 and 97%) and oral (63 and 60%) administration of methylparaoxon, the gastrointestinal absorption seemed to be about 60%. It was assumed, that the kinetics were linear in this dose range.

The concentration of the main metabolites paranitrophenol (PNP) and dimethylphosphate (DMP) in the urine of 4 human volunteers, following 2 days of ingestion of 2 or 4 mg methyl parathion, is shown in Table 17. Unmetabolized traces of methyl parathion were also found in the urine, which was collected after 4-, 8-, and 24-h (Morgan et al., 1977). The urinary excretion of nitrophenol was 60% within 4 h, 86% within 8 h, and approximately 100% within 24 h following ingestion. Table 17 shows the dependence of urinary metabolite excretion on methyl parathion dosage.

Table 17. *p*-Nitrophenol (PNP) and dimethylphosphate (DMP) concentrations in 24-h urine samples collected from human volunteers following administration of 2 or 4 mg methyl parathion[a]

	PNP		DMP	
	mean	range	mean	range
a) **2 mg methyl parathion**				
urinary concentration (mg/litre)	0.13	0.08-0.20	0.06	0.02-0.11
24-h excretion (mg)	0.29	0.14-0.43	0.12	0.07-0.16
excretion per g creatinine (mg/g)	0.16	0.10-0.23	0.06	0.03-0.10
b) **4 mg methyl parathion**				
urinary concentration (mg/litre)	0.34	0.16-0.61	0.14	0.05-0.23
24-h excretion (mg)	0.58	0.34-0.88	0.23	0.12-0.41
excretion per g creatinine (mg/g)	0.31	0.15-0.42	0.13	0.06-0.20

[a] Adapted from: Morgan et al. (1977).

6.5 Retention and turnover

Braeckman et al. (1980) injected 10 mg methyl parathion per kg body weight intravenously into dogs, recorded the uptake of methyl parathion, and determined a harmonic mean terminal half-life of 7.2 h. Five hours after the injection, the concentration decreased to 30% of the initial value. Primarily, the peripheral body compartments contained this residual methyl parathion. The excretion was completed within 35 h.

The velocity of the excretion of the main metabolites after oral or intravenous application was similar. However, the bioavaibility after oral intake was reduced by first-pass extraction by the liver compared with the intravenous application. Methyl parathion was shown to bind to a great extent (90%) to plasma proteins in both dogs and humans (Braeckman et al., 1983).

7. EFFECTS ON ORGANISMS IN THE ENVIRONMENT

7.1 Microorganisms

7.1.1 Bacteria and fungi

Soil concentrations of methyl parathion of 5 mg/kg or more were found to reduce microbial reductive potential (Reddy & Gambrell, 1985).

In biotests for sanitary control of water samples, growth inhibition of *Escherichia coli* by several toxicants was studied in a liquid medium (Vogel-Bonner medium, supplemented with thymine and glucose at 37 °C). The minimal concentrations of methyl parathion that significantly increased growth rate and doubling time of *E. coli* were reported to be 62.5 mg/litre and 125 mg/litre; the bacterium used the compound as a carbon source (Espigares et al., 1990).

Portier et al. (1983) tested the effects of methyl parathion (1.5 or 5 mg/litre) on the reproduction of aquatic microorganisms from drainage basins in laboratory experiments, using static or flow-through approaches (28 °C; pH 7.5; 22%; 22 days; or 28 °C; pH 7.2; 0%; 24 days). In bacteria and *Actinomycetes*, methyl parathion had a positive effect on the development. In fungi and yeasts, slight negative effects were found that were related to the test conditions rather than to the toxicant concentration. In general, a concentration of up to 5 mg methyl parathion/litre resulted in increased activity and biomass production in a microbial community, being used as carbon source by the microorganisms (Portier & Meiers, 1982).

Bhunia et al. (1991) cultured *Nostoc muscorum*, a blue-green alga (*Cyanobacterium*), which is a major nitrogen-fixing organism in tropical soil, with methyl parathion at 5, 10, 20, or 35 mg/litre. Only the highest concentration significantly reduced the growth of the cells in culture. However, the chlorophyll-*a* contents of the cultures were marginally reduced at 5 mg methyl parathion/litre and substantially reduced at 10 mg/litre. Nitrogenase activity was reduced to < 50% of control levels at 10 mg/litre.

7.1.2 Algae

The 96-h EC_{50}, i.e., the calculated concentration of methyl parathion that would inhibit growth by 50% of the diatom *Skeletonema costatum*, ranged between 5.0 and 5.3 mg/litre (Walsh & Alexander, 1980; Walsh et al., 1987).

Exposure of cultures of *Chlorella protothecoides* to 26-80 µg methyl parathion/litre resulted in decreases in cell growth, as measured by cell count, and chlorophyll and protein contents (Saroja-Subbaraj & Bose, 1982; Saroja-Subbaraj & Bose, 1983a). These effects were correlated with a reduction in photosynthetic electron transfer (Saroja-Subbaraj & Bose, 1983a; Saroja-Subbaraj & Bose, 1983b). Recovery from the effect on photosynthesis occurred after removal of the pesticide. Tolerance to the effect of methyl parathion on cell growth occurred for several weeks after exposure (Saroja-Subbaraj & Bose, 1984).

In a natural phytoplankton community, addition of 1 mg methyl parathion/litre led to a 5% decrease in the productivity (Butler, 1964).

An algal bloom (species not specified) in a methyl parathion-treated pond was suggested to have been induced by the mortality of herbivorous mayfly larvae and *Daphnia* (Crossland & Elgar, 1983).

7.2 Aquatic animals

The acute effects of methyl parathion on aquatic animals in laboratory studies are presented in Table 18. The data show that the sensitivity of aquatic animals to methyl parathion varies considerably between species.

LC_{50} values of more than 1 mg/litre have been found for some freshwater biota (molluscs, fish, and amphibians). Insect sensitivity to methyl parathion depends not only on the species but on the life stage. In general, instar I larvae are more affected than instar IV larvae. Apperson et al. (1978) showed that larvae may develop a resistance to methyl parathion. Both freshwater and marine crustaceans are sensitive to methyl parathion with EC_{50} values ranging from 0.002 to 0.050 mg/litre. In general, copepods were less sensitive than decapods in laboratory tests.

Table 18. Acute effects of methyl parathion on aquatic animals in laboratory studies

Species	Life stage	Test period (h)	Experimental conditions	Criterion effect measured[a]	Concentration (µg/litre)	Remarks[b]	References
MOLLUSCA							
Freshwater mussel							
Lamellidens marginalis		48	st.[c]	m, LC$_{50}$	20 000	-	Moorthy et al. (1983)
Lamellidens marginalis		48	st.	m, LC$_{50}$	25 000	-	Moorthy et al. (1983)
Lamellidens marginalis	20 g	48	st.	m, LC$_{50}$	23 400	-	Rao et al. (1983)
Eastern oyster							
Crassostrea virginica	larvae	48	st.; natural seawater 25 °C	d, EC$_{50}$	12 000	P: 99% s: TEG	Mayer (1987)
Marine hard clam							
Mercenaria mercenaria	adult	96	St.; wellwater; 24°/oo [d] 20 °C pH8	no effect	25 000	s: acetone	Mayer (1987)
Nassa dosoleta	adult	96	st.; wellwater; 24°/oo [d] 20 °C pH8	no effect	25 000	s: acetone	Mayer (1987)

Table 18 (continued)

ANNELIDA (Estuarine)

Species	Stage	Duration (h)	Conditions	Endpoint	Value	P/s	Reference
Branchiura sowerbyi	-	72	st.; 4.4 °C	m, 100%	4000	P: techn.gr. s: acetone	Naqvi (1973)
Branchiura sowerbyi	-	72	st.; 21 °C	m, 0%	4000	P: techn.gr. s: acetone	Naqvi (1973)
Branchiura sowerbyi	-	72	st.; 32.2 °C	m, 100%	4000	P: techn.gr. s: acetone	Naqvi (1973)

CRUSTACEA (Freshwater)

Water flea

Species	Stage	Duration (h)	Conditions	Endpoint	Value	P/s	Reference
Daphnia longispira	adult	24	st.; dechlorinated tap-water; 19.5 °C; H 250[e]	i, LC_{50}	2.4	P: 93.8% s: acetone	Stephenson & Kane (1984)
Daphnia magna	< 24 h old	24	st.; dechlorinated tap-water; 19.5 °C; H 250	i, LC_{50}	4.1	P: 93.8% s: acetone	Stephenson & Kane (1984)
Daphnia magna	adult	24	-	i, LC_{50}	5.4	P: 93.8% s: acetone	Stephenson & Kane (1984)
Daphnia magna	< 24 h old	48	st.; artificial water; 18 °C	i, LC_{50}	7.8-9.1	P: 99% s: acetone	Dortland (1980)

Table 18 (continued)

Species	Life stage	Test period (h)	Experimental conditions	Criterion effect measured[a]	Concentration (µg/litre)	Remarks[b]	References
Daphnia magnia	first instar	48	st.; reconst. water; 21 °C pH 7.2-7.5; H40-50	i, LC$_{50}$	0.14	P: 98.7 s: acetone	Mayer & Ellersieck (1986)
Daphnia pulex	-	3	st.; 24 °C	m, LC$_{50}$	8.5	-	Nishiuchi & Hashimoto, (1967)
Moira macrocopa	-	3	st.; 24 °C	m, LC$_{50}$	5.5	-	Nishiuchi & Hashimoto (1967)
Simocephalus secrultus	first instar larva	48	st.; reconst. water; 15 °C pH 7.2-7.5 H 40-50	i, LC$_{50}$	0.37	P: 98.7% s: acetone	Mayer & Ellersieck (1986)
Scud							
Gammarus fasciatus	adult	96	st.; reconst. water; 15 °C pH 7.2-7.5; H 40-50	m, LC$_{50}$	3.8	P: 98.7% s: acetone	Mayer & Ellersieck (1986)

Table 18 (continued)

Field crab

		(h)	conditions	method	value	purity/solvent	reference
Oziotelphusa senex senex	-	48	st.: tap-water; pH 7.3; 30 °C DO 6.2[f]; H 38	m, LC$_{50}$	1000	P: techn.gr. s: acetone	Reddy et al. (1986a)

Crayfish

Orconectes nais	adult	96	st.: reconst. water; 15 °C pH 7.2-7.5; H 162-272	m, LC$_{50}$	15	P: 98.7% s: acetone	Mayer & Ellersieck (1986)
Procambarus acutus	2.5-3.5 cm	96	st.: tap-water; pH 8.4; H 100	m, LC$_{50}$	3	P: techn.gr.	Cheah et al. (1980)
from clean area	1.2-1.5 cm	48	st.: tap-water; pH 8.7; H 10	m, LC$_{50}$	2.4	P: techn.gr. s: acetone	Albaugh (1972)
from treated area	1.2-1.5 cm	48	st.: tap-water; pH 8.7; H 10	m, LC$_{50}$	3.4	P: techn.gr. s: acetone	Albaugh (1972)
Procambarus clarkii	8.9 cm	36	st.: distilled water 22.2-25-5 °C	m, LC$_{50}$	41	P: 51%	Chang & Lange (1967)
Procambarus clarkii	8.9 cm	24	st.: tap-water 16-32 °C pH 7.6	m, LC$_{50}$	50	P: tech.gr.	Muncy & Oliver (1963)

Table 18 (continued)

Species	Life stage	Test period (h)	Experimental conditions	Criterion effect measured[a]	Concentration (µg/litre)	Remarks[b]	References
Procambarus clarkii	8.9 cm	48	st.; tap-water 16-32 °C pH 7.6	m, LC$_{50}$	40	P: tech.gr	Muncy & Oliver (1963)
Procambarus clarkii	8.9 cm	72	st.; tap-water 16-32 °C pH 7.6	m, LC$_{50}$	40	P: tech.gr.	Muncy & Oliver (1963)
ESTUARINE AND MARINE							
Copepod							
Acartia tonsa	-	96	st.; natural seawater; 22°/oo 22 °C; pH 8.1-8.2; DO 7-7.6	m, LC$_{50}$	28	P: 99% s: TEG	Mayer (1987)
Acartia tonsa	adult	96	st.; synthetic seawater; 22°/oo 17 °C	m, LC$_{50}$	890	P: 80%	Khattat & Farley (1976)

Table 18 (continued)

Sand shrimp

Species		Duration (h)	Conditions	Effect	Value	Substance	Reference
Crangon septemspinosa	2.6 cm 0.25 g	24	st.; wellwater 24°/oo; 20 °C; pH 8 DO 7.1-7.7	m, LC$_{50}$	11	s: acetone	Eisler (1969)
Crangon septemspinosa	2.6 cm 0.25 g	48	st.; wellwater 24°/oo; 20 °C; pH 8 DO 7.1-7.7	m, LC$_{50}$	3	s: acetone	Eisler (1969)
Crangon septemspinosa	2.6 cm 0.25 g	96	st.; wellwater 24°/oo; 20 °C; pH 8 DO 7.1-7.7	m, LC$_{50}$	2	s: acetone	Eisler (1969)

Mysid shrimp

Species		Duration (h)	Conditions	Effect	Value	Substance	Reference
Mysidopsis bahia	24 h old	96	st.; natural seawater; 20°/oo; 25 °C; DO 4.3-5.5	m, LC$_{50}$	0.98	P: 99% s: TEG	Mayer (1987)
Mysidopsis bahia	24 h old	96	st.; natural seawater; 20°/oo; 25 °C; DO 4.3-5.5	no effect	0.32	P: 99% s: TEG	Mayer (1987)
Mysidopsis bahia	24 h old	96	flow-through natural seawater; 20°/oo; 25 °C; DO 4.3-5.5	m, LC$_{50}$	0.77	P: 99% s: TEG	Mayer (1987)

Table 18 (continued)

Species	Life stage	Test period (h)	Experimental conditions	Criterion effect measured[a]	Concentration (μg/litre)	Remarks[b]	References
Mysidopsis bahia	< 24 h old	96	flow-through 14°/oo; 19.5 °C;	m, LC$_{50}$	0.78	P: 99% s: TEG	Mayer (1987)
Mysidopsis bahia	juvenile	96	flow-through 22-28 °C	m, LC$_{50}$	0.77	s: TEG	Nimmo et al. (1981)
Mysidopsis bahia	juvenile	96	flow-through 22-28 °C	MATC[g]	0.11-0.16	s: TEG	Nimmo et al. (1981)
Hermit crab							
Pagurus longicarpus	3.5 mm 0.28 g	24	st.; wellwater 24°/oo; 20 °C; ph 8: DO 7.1-7.7	m, LC$_{50}$	23	s: acetone	Eisler (1969)
Pagurus longicarpus	3.5 mm 0.28 g	48	st.; wellwater 24°/oo; 20 °C; ph 8: DO 7.1-7.7	m, LC$_{50}$	7	s: acetone	Eisler (1969)
Pagurus longicarpus	3.5 mm 0.28 g	96	st.; wellwater 24°/oo; 20 °C; ph 8: DO 7.1-7.7	m, LC$_{50}$	7	s: acetone	Eisler (1969)

Table 18 (continued)

Crab							
Portunus trit-uberculatus	Zoöe IV stage	24	25 °C	m, LC_{50}	0.17-0.5	-	Hirayama & Tamaoi (1980)
Grass shrimp							
Palaemonetes vulgaris	31 mm 0.47 g	24	st.; wellwater; 24°/oo; 20 °C; ph 8; DO 7.1-7.7	m, LC_{50}	15	s: acetone	Eisler (1969)
Palaemonetes vulgaris	31 mm 0.47 g	48	st.; wellwater; 24°/oo; 20 °C; ph 8; DO 7.1-7.7	m, LC_{50}	10	s: acetone	Eisler (1969)
Palaemonetes vulgaris	31 mm 0.47 g	96	st.; wellwater; 24°/oo; 20 °C; ph 8; DO 7.1-7.7	m, LC_{50}	3	s: acetone	Eisler (1969)
Brown shrimp							
Penaeus aztecus	adult	24	flow-through 29°/oo; 25 °C	m, LC_{50}	5.5	s: acetone	Butler (1964)
Penaeus aztecus	adult	48	flow-through 29°/oo; 25 °C	m, LC_{50}	5.5	s: acetone	Butler (1964)

Table 18 (continued)

Species	Life stage	Test period (h)	Experimental conditions	Criterion effect measured[a]	Concentration (μg/litre)	Remarks[b]	References
Pink shrimp							
Penaeus duorarum	-	-	flow-through 17-31 °/oo 7.6-28.8 °C	m, LC$_{50}$	1.9	s: acetone + TEG	Schoor & Brausch, (1980)
Penaeus duorarum	post-larvae	96	flow-through natural seawater 20 °/oo; 25 °C	m, LC$_{50}$	1.2	s: TEG P: 99%	Mayer (1987)
Japanese shrimp							
Penaeus japonicus	post-larve	24	25 °C	m, LC$_{50}$	0.5-0.9	-	Hirayama & Tamaoi (1980)
Shrimp							
Penaeus stylirostris	post-larvae	96	st.; natural seawater 20 °/oo; 25 °C; DO 5.6-6.3	m, LC$_{50}$	1.4	s: TEG P: 99%	Mayer (1987)
Penaeus monodon	adult	96	st.; 15 °/oo; 23 °C; pH 7.3	m, LC$_{50}$	148	-	Reddy & Rao (1986)
Penaeus indicus	adult	96	st.; 15 °/oo; 23 °C; pH 7.3	m, LC$_{50}$	98	-	Reddy & Rao (1986)

Table 18 (continued)

Species	Stage	Duration (h)	Conditions	Method	Value	Purity/Solvent	Reference
Penaeus indicus	(inter-molt) 2.5 g	48	st.; seawater; 15 °/oo; 23 °C pH 7.1	m, LC$_{50}$	95	-	Reddy & Rao (1986)
Metapenaeus monoceros	adult	96	st.; 15 °/oo 23 °C; pH 7.3	m, LC$_{50}$	102	-	Reddy & Rao (1986)
Metapenaeus monoceros	(inter-molt) 2.5 g	48	st.; seawater; 15 °/oo; 23 °C; pH 7.1	m, LC$_{50}$	120	-	Reddy & Rao (1988)
Metapenaeus dopsoni	adult	96	st.; 15 °/oo 23 °C; pH 7.3	m, LC$_{50}$	115	-	Reddy & Rao (1986)
INSECTA							
Mosquito							
Culex pipiens	4th instar larva	24	st.; 28 °C; deionized water	m, LC$_{50}$	30	P: 98.2% s: ethanol	Yasuno et al. (1965)
Culex pipiens	4th instar larva	24	st.; 28 °C; polluted water	m, LC$_{50}$	2000	P: 98.2% s: ethanol	Yasuno et al. (1965)
Culex pipiens	4th instar larva	96	st.; 28 °C; deionized water	m, LC$_{50}$	30	P: 98.2% s: ethanol	Yasuno et al. (1965)

Table 18 (continued)

Species	Life stage	Test period (h)	Experimental conditions	Criterion effect measured[a]	Concentration (µg/litre)	Remarks[b]	References
Culex pipiens	4th instar larva	96	st.; 28 °C; polluted water	m, LC₅₀	80 000	P: 98.2% s: ethanol	Yasuno et al. (1965)
Chaetorus astictopus	lst instar larva	24	st.; lake water; 25 °C	m, LC₅₀	1.6	P: techn.gr. s: acetone (1962 exper.)	Apperson et al. (1978)
Chaetorus astictopus	4th instar larva	24	st.; lake water; 25 °C	m, LC₅₀	30	P: techn.gr. s: acetone (1962 exper.)	Apperson et al. (1978)
Chaetorus astictopus	1st instar larva	24	st.; lake water; 25 °C	m, LC₅₀	18	P: techn.gr. s: acetone (1978 exper.)	Apperson et al. (1978)
Chaetorus astictopus	4th instar larva	24	st.; lake water; 25 °C	m, LC₅₀	85	P: techn.gr. s: acetone (1978 exper.)	Apperson et al. (1978)

Table 18 (continued)

Damselfly							
Ischnura verticalus	larva	96	st.; reconst. water; 15 °C pH 7.2-7.5 H 167-272	m, LC$_{50}$	33	P: 98.7% s: acetone	Mayer & Ellersieck (1987)
FISH (Freshwater)							
Betta splendens	adult	120	tap-water; 25 °C pH 7-7.4	m, LC$_{50}$	7500-8000	s: hexane	Walsh & Hanselka (1972
Goldfish							
Carassius auratus	0.6-1.7 g	96	st.; reconst. water; 18 °C; pH 7.1	m, LC$_{50}$	9000	P: 80% s: acetone	Mayer & Ellersieck (1986)
Carassius auratus	4.6 cm 1.2 g	24	st.; dest. water; 25 °C; pH 7.4-7.5 H 20; DO 4-8	m, LC$_{50}$	14 000	P: 80% s: acetone	Pickering et al. (1962)
Carassius auratus	4.6 cm 1.2 g	48	st.; distilled water; 25 °C; pH 7.4-7.5 H 20; DO 4-8	m, LC$_{50}$	12 000	P: 80% s: acetone	Pickering et al. (1962)

Table 18 (continued)

Species	Life stage	Test period (h)	Experimental conditions	Criterion effect measured[a]	Concentration (µg/litre)	Remarks[b]	References
Carassius auratus	4.6 cm 1.2 g	96	st.; distilled water; 25 °C; pH 7.4-7.5 H 20; DO 4-8	m, LC_{50}	12 000	P: 80% s: acetone	Pickering et al. (1962)
Golden carp							
Cyprinus auratus	-	48	st.; 24 °C	m, LC_{50}	> 10 000	P: 80% s: acetone	Nishiuchi & Hashimoto (1967)
Carp							
Cyprinus carpio	< 1 year	24	st.; 20 °C pH 7.2; DO 6; H 50	m, LC_{50}	27 600	P: 80% s: acetone	Rehwoldt et al. (1977)
Cyprinus carpio	< 1 year	48	st.; 20 °C pH 7.2; DO 6; H 50	m, LC_{50}	21 200	P: 80% s: acetone	Rehwoldt et al. (1977)
Cyprinus carpio	< 1 year	96	st.; 20 °C pH 7.2; DO 6; H 50	m, LC_{50}	14 800	P: 80% s: acetone	Rehwoldt et al. (1977)

Table 18(continued)

Species		Duration (h)	Conditions		Concentration	P/s	Reference
Cyprinus carpio	35 g	48	st.; 20 °C pH 7.2; DO 6; H 50	m, LC_{50}	12 000	P: 80% s: acetone	Nagaratnamma & Ramamurthi (1982)
Cyprinus carpio	0.6–1.7 g	96	st.; reconst. water; 18 °C pH 7.1	m, LC_{50}	7130	P: 80% s: acetone	Mayer & Ellersieck (1986)
Cyprinus carpio	0.6–1.7 g	96	st.; reconst. water; 18 °C pH 7.2-7.5 H 40-50	m, LC_{50}	8900	P: techn.gr. s: acetone	Johnson & Finley (1980)
Cyprinus carpio	0.6 g	48	st.; 24 °C	m, LC_{50}	> 10 000	P: techn.gr. s: acetone	Nishiuchi & Hashimoto (1967)
Banded killifish							
Fundulus diaplanus	< 1 year	24	st.; 20 °C pH 7.2; DO 6; H 50	m, LC_{50}	24 900	P: techn.gr. s: acetone	Rehwoldt et al. (1977)
Fundulus diaplanus	< 1 year	48	st.; 20 °C pH 7.2; DO 6; H 50	m, LC_{50}	18 600	P: techn.gr. s: acetone	Rehwoldt et al. (1977)
Fundulus diaplanus	< 1 year	96	st.; 20 °C pH 7.2; DO 6; H 50	m, LC_{50}	15 200	P: techn.gr. s: acetone	Rehwoldt et al. (1977)

115

Table 18 (continued)

Species	Life stage	Test period (h)	Experimental conditions	Criterion effect measured[a]	Concentration (μg/litre)	Remarks[b]	References
Mosquito fish							
Gambusia affinis	adult non re-sistent	48	st.; dechlorinated tap-water	m, LC_{50}	13 480	P: 99% s: methoxy-ethanol	Chambers & Yarbrough (1974)
Gambusia affinis	adult non re-sistent	48	st.; dechlorinated tap-water	m, LC_{50}	17 480	P: 99% s: methoxy-ethanol	Chambers & Yarbrough (1974)
Catfish							
Heteropneustes fossilis	adult (fem)	96	24 °C; pH 7.7; DO 6; H 117	m, LC_{50}	7000	s: acetone	Srivastava & Singh (1981)
Heteropneustes fossilis	16 cm 35 g	24	st.; 23 °C; pH 7.7; DO 6.1; H 115	m, LC_{50}	9400	s: acetone	Singh & Srivastava (1982)
Heteropneustes fossilis	16 cm 35 g	48	st.; 23 °C; pH 7.7; DO 6.1; H 115	m, LC_{50}	8600	s: acetone	Singh & Srivastava (1982)
Heteropneustes fossilis	16 cm 35 g	72	st.; 23 °C; pH 7.7; DO 6.1; H 115	m, LC_{50}	8000	s: acetone	Singh & Srivastava (1982)

Table 18 (continued)

Species		Duration (h)	Conditions		Value		Reference
Heteropneustes fossilis	16 cm 35 g	96	st.; 23 °C; pH 7.7; DO 6.1; H 115	m, LC_{50}	7000	s: acetone	Singh & Srivastava (1982)
Black Bullhead							
Ictalurus melas	0.6-1.7 g	96	st.; reconst. water; 18 °C pH 7.1	m, LC_{50}	6640	P: 80% s: acetone	Mayer & Ellersieck (1986)
Catfish							
Mystus cavasius	6-8 cm 7 g	96	26-30 °C	m, LC_{50}	5900	-	Murty & Ramani (1982)
Channel Catfish							
Ictalurus punctatus	1.4 g	96	st.; reconst. water; 18 °C pH 7.2-7.5; H 40-50	m, LC_{50}	5240	P. techn.gr. s: acetone	Mayer & Ellersieck (1986)
Guppy (*Poecilia reticulata*)							
Lebistes reticulatus	6 mon.	24	st.; distilled water; 25 °C pH 7.4-7.5; H 20, DO 4-8	m, LC_{50}	11 000	P. 80% s: acetone	Pickering et al. (1962)

Table 18 (continued)

Species	Life stage	Test period (h)	Experimental conditions	Criterion effect measured[a]	Concentration (μg/litre)	Remarks[b]	References
Lebistes reticulatus	6 mon.	48	st.; distilled water; 25 °C pH 7.4-7.5; H 20, DO 4-8	m, LC_{50}	9800	P. 80% s: acetone	Pickering et al. (1962)
Lebistes reticulatus	6 mon.	96	st.; distilled water; 25 °C pH 7.4-7.5; H 20, DO 4-8	m, LC_{50}	9800	P. 80% s: acetone	Pickering et al. (1962)
Lebistes reticulatus	< 1 year	24	st.; 20 °C pH 7.2; DO 6 H 20	m, LC_{50}	12 200	P. 80% s: acetone	Rehwoldt et al. (1977)
Lebistes reticulatus	< 1 year	48	st.; 20 °C pH 7.2; DO 6 H 20	m, LC_{50}	9400	P. 80% s: acetone	Rehwoldt et al. (1977)
Lebistes reticulatus	< 1 year	96	st.; 20 °C pH 7.2; DO 6 H 20	m, LC_{50}	6200	P. 80% s: acetone	Rehwoldt et al. (1977)

Table 18 (continued)

Green sunfish

Lepomis cyanellus	0.8 g	96	st.; reconst. water; 17 °C pH 7.2-7.5; H 40-50	m, LC_{50}	6860	P: techn.gr. s: acetone	Mayer (1987)
Lepomis cyanellus	0.8 g	48	st.; tap-water 20 °C	m, LC_{50}	> 5000	P: techn.gr. s: acetone	Minchew & Ferguson (1969)
Pumpkinseed							
Lepomis gibbosus	40-50 g	24	injection, st.	m, LD_{50}	> 2500	P: 99% s: corn oil	Benke et al. (1974)
Lepomis gibbosus	< 1 year	24	st.: 20 °C pH 7.2; DO 6; H 50	m, LD_{50}	4900	P: 99% s: corn oil	Rehwoldt et al. (1977)
Lepomis gibbosus	< 1 year	48	st.: 20 °C pH 7.2; DO 6; H 50	m, LD_{50}	3600	P: 99% s: corn oil	Rehwoldt et al. (1977)
Lepomis gibbosus	< 1 year	96	st.: 20 °C pH 7.2; DO 6; H 50	m, LD_{50}	3600	P: 99% s: corn oil	Rehwoldt et al. (1977)

Table 18 (continued)

Species	Life stage	Test period (h)	Experimental conditions	Criterion effect measured[a]	Concentration (µg/litre)	Remarks[b]	References
Bluegill sunfish							
Lepomis macrochirus	finger-ling	24	st.; reconst. water; 18 °C pH 7; H 17	m, LC_{50}	6470	P: 44.6% s: water	McCann & Jasper (1972)
Lepomis macrochirus	0.6-1.7 g	96	st.; reconst. water; 18 °C pH 7.1	m, LC_{50}	5720	P: 80% s: acetone	Macek & McAllister (1970)
Lepomis macrochirus	4-6 cm 1.2 g	24	st.; distilled water; 25 °C pH 7.4-7.5 H 20; DO 4-8	m, LC_{50}	9800	P: 80% s: acetone	Pickering et al. (1962)
Lepomis macrochirus	4-6 cm 1.2 g	48	st.; distilled water; 25 °C pH 7.4-7.5 H 20; DO 4-8	m, LC_{50}	8600	P: 80% s: acetone	Pickering et al. (1962)
Lepomis macrochirus	4-6 cm 1.2 g	96	st.; distilled water; 25 °C pH 7.2-7.5; H 20; DO 4-8	m, LC_{50}	2400	P: 80% s: acetone	Pickering et al. (1962)

Table 18 (continued)

Lepomis macrochirus	1 g	96	st.; reconst. water; 17 °C pH 7.2-7.5 H 40-50	m, LC$_{50}$	4380	P: techn.gr. s: acetone	Mayer & Ellersieck (1986)
Lepomis macrochirus	0.6-1.7 g	96	st.; reconst. water; 18 °C pH 7.1	m, LC$_{50}$	5170	P: 80% s: acetone	Macek & McAllister (1970)
Largemouth bass							
Micropterus salmoides	0.6-1.7 g	96	st.; reconst. water; 18 °C pH 7.1	m, LC$_{50}$	5220	P: 80% s: acetone	Mayer & Ellersieck (1986)
Mystus cavasius	-	96	-	m, LC$_{50}$	5900	-	Murty & Ramani (1982)
Golden shiner							
Notemigonus chrysoleuces	-	48	st.; tap-water: 20 °C	m, LC$_{50}$	> 5000	P: techn.gr. s: acetone	Minchew & Ferguson (1969)
Coho salmon							
Oncorhynchus kisutch	0.6-1.7 g	96	st.; reconst. water; 13 °C pH 7.1	m, LC$_{50}$	5300	P: 80% s: acetone	Mayer & Ellersieck (1986)

Table 18 (continued)

Species	Life stage	Test period (h)	Experimental conditions	Criterion effect measured[a]	Concentration (μg/litre)	Remarks[b]	References
Medaka							
Oryzias latipes	-	48	st.; 24 °C	m, LC_{50}	7500	P: techn.gr. s: acetone	Nishiuchi & Hashimoto (1967)
Yellow perch							
Perca flavescens	1.4 g	96	st.; reconst. water; 18 °C pH 7.2-7.5; H 40-50	m, LC_{50}	3060	P: techn.gr. s: acetone	Mayer & Ellersieck (1986)
Punti							
Puntius puckelli	6-8.5 cm	24	st.; 27.9 °C pH 8.3; H 130	m, LC_{50}	2900	P: 50%	Rao et al. (1967)
Puntius puckelli	6-8.5 cm	48	st.; 27.9 °C pH 8.3; H 130	m, LC_{50}	2700	P: 50%	Rao et al. (1967)
Puntius puckelli	6-8.5 cm	96	st.; 27.9 °C pH 8.3; H 130	m, LC_{50}	2100	P: 50%	Rao et al. (1967)

Table 18 (continued)

Fathead minnow

Species		Duration (h)	Conditions		Value	Purity/solvent	Reference
Pimephales promelas	1.2 g	96	st.; reconst. water; 18 °C pH 7.2-7.5; H40-50	m, LC_{50}	8300	P: techn.gr. s: acetone	Mayer & Ellersieck (1986)
Pimephales promelas	-	48	flow-through	m, LC_{50}	7400	P: 98.5% s: acetone	Solon & Nair (1970)
Pimephales promelas	-	96	flow-through	m, LC_{50}	3750	P: 98.5% s: acetone	Solon & Nair (1970)
Pimephales promelas	newly hatched larvae	96	st.; sterilized water; 25 °C pH 7.4-7.8; DO 6.5-8.4; H 64	m, LC_{50}	4460	P: 80%	Jarvinen & Tanner (1982)
Pimephales promelas	newly hatched larvae	96	st.; sterilized water; 25 °C pH 7.4-7.8;	m, LC_{50} DO 6.5-8.4; H 64	1220	P: 80% stock solution aged 11 weeks	Jarvinen & Tanner (1982)
Pimephales promelas	newly hatched larvae	96	st.; sterilized water; 25 °C pH 7.4-7.8; DO 6.5-8.4; H 64	m, LC_{50}	8170	P: 80% controlled-release formulation	Jarvinen & Tanner (1982)

Table 18 (continued)

Species	Life stage	Test period (h)	Experimental conditions	Criterion effect measured[a]	Concentration (µg/litre)	Remarks[b]	References
Pimephales promelas	newly hatched larvae	96	st.; sterilized water; 25 °C pH 7.4-7.8; DO 6.5-8.4; H 64	m, LC_{50}	3470	P: 80% controlled release formulation for 11 weeks	Jarvinen & Tanner (1982)
Pimephales promelas	newly hatched larvae	96	flow-through; sterilized water 25 °C; pH 7.4-7.8; DO 6.5-8.4; H 64	m, LC_{50}	5360	P: 80%	Jarvinen & Tanner (1982)
Pimephales promelas	newly hatched larvae	96	flow-through; sterilized water 25 °C; pH 7.4-7.8; DO 6.5-8.4; H 64	m, LC_{50}	6910	P: 80% controlled release formulation	Jarvinen & Tanner (1982)
Pimephales promelas	4-6 cm 1-2 g	24	st.; distilled water; 25 °C pH 7.4-7.5; H 20; DO 4-8	m, LC_{50}	13 000	P: 80% s: acetone	Pickering et al. (1962)

124

Table 18 (continued)

Pimephales promelas	4-6 cm 1-2 g	48	st.; distilled water; 25 °C pH 7.4-7.5; H 20; DO 4-8	m, LC$_{50}$	9800	P: 80% s: acetone	Pickering et al. (1962)
Pimephales promelas	4-6 cm 1-2 g	96	st.; distilled water; 25 °C pH 7.4-7.5; H 20; DO 4-8	m, LC$_{50}$	9500	P: 80% s: acetone	Pickering et al. (1962)
White perch							
Roccus americanus	< 1 year	24	st.; 12 °C	m, LC$_{50}$ pH 7.2	22 400	P: 80% s: acetone	Rehwoldt et al. (1977)
Roccus americanus	< 1 year	48	st.; 12 °C pH 7.2	m, LC$_{50}$	18 600	P: 80% s: acetone	Rehwoldt et al. (1977)
Roccus americanus	< 1 year	96	st.; 12 °C pH 7.2	m, LC$_{50}$	14 000	P: 80% s: acetone	Rehwoldt et al. (1977)
Cutthroat trout							
Salmo clarki	0.2 g	96	st.; reconst. water; 12 °C pH 7.2-7.5; H 162-272	m, LC$_{50}$	1850	P: techn.gr. s: acetone	Mayer & Ellersieck (1986)

Table 18 (continued)

Species	Life stage	Test period (h)	Experimental conditions	Criterion effect measured[a]	Concen- tration (µg/litre)	Remarks[b]	References
Rainbow trout (Oncorhynchus mykiss)							
Salmo gairdneri	1.1 g	96	st.; reconst. water; 12 °C; pH 7.2-7.5; H 162-272	m, LC$_{50}$	3700	P: techn.gr. s: acetone	Mayer & Ellersieck (1986)
Salmo gairdneri	0.6- 1.7 g	96	st.; reconst. water; 13 °C; pH 7.1	m, LC$_{50}$	2750	P: 80% s: acetone	Macek & McAllister (1970)
Salmo gairdneri	24 mm	96	st.; 12 °C	m, LC$_{50}$	2800	P: 76.8%	Palawski et al. (1983)
Brown trout							
Salmo trutta	0.6- 1.7 g	96	st.; reconst. pH 7.1 water; 13 °C	m, LC$_{50}$	4750	P: 80% s: acetone	Mayer & Ellersieck (1986)

Table 18 (continued)

		Duration (h)	Conditions		Concentration	Preparation/solvent	Reference
Brook trout							
Salvelinus fontinalis	0.5 g	96	st.; reconst. water; 12 °C; pH 7.2-7.5; H 40-50	m, LC_{50}	3780	P: techn.gr. s: acetone	**Mayer & Ellersieck (1986)**
Northern pike							
Esox lucius	0.4 g	24	st.; 18 °C pH 7.1; H 44	m, LC_{50}	760	P: techn.gr. s: acetone	**Mayer & Ellersieck (1986)**
Tilapia							
tilapia mossambica	-	48	st.; 26-28 °C pH 7; H 140	m, LC_{50}	266	P: techn.gr. s: 2-meth-oxyethanol	Rao & Rao (1983)
ESTUARINE AND MARINE							
American eel							
Anguilla rostrata	59 mm- 0.14 g	24	st.; underground wellwater; 24‰; 20 °C; pH 8; DO 7.1-7.7	m, LC_{50}	27 600	P: act. ingredient s: acetone	Eisler (1970a)

Table 18 (continued)

Species	Life stage	Test period (h)	Experimental conditions	Criterion effect measured[a]	Concentration (µg/litre)	Remarks[b]	References
Anguilla rostrata	59 mm-0.14 g	48	st.; underground wellwater; 24°/oo; 20 °C; pH 8; DO 7.1-7.7	m, LC$_{50}$	22 400	P: act. ingredient s: acetone	Eisler (1970a)
Anguilla rostrata	59 mm-0.14 g	96	st.; underground wellwater; 24°/oo; 20 °C; pH 8; DO 7.1-7.7	m, LC$_{50}$	16 900	P: act. ingredient s: acetone	Eisler (1970a)
Anguilla rostrata	< 1 year	24	st.; 20 °C. pH 7.2; DO 6; H 50	m, LC$_{50}$	42 600	P: act. ingredient s: acetone	Rehwoldt et al. (1977)
Anguilla rostrata	< 1 year	48	st.; 20 °C. pH 7.2; DO 6; H 50	m, LC$_{50}$	37 200	P: act. ingredient s: acetone	Rehwoldt et al. (1977)
Anguilla rostrata	< 1 year	96	st.; 20 °C. pH 7.2; DO 6; H 50	m, LC$_{50}$	6300	P: act. ingredient s: acetone	Rehwoldt et al. (1977)

Table 18 (continued)

Sheepshead minnow

Cyprinodon variegatus	28 days old	96	st.; natural seawater; 20°/oo; 25 °C; DO 4.6-5.7	m, LC$_{50}$	12 000	P: 99% s: TEG	Mayer (1987)
Cyprinodon variegatus	28 days old	96	st.; natural seawater; 20°/oo; 25 °C; DO 4.6-5.7	no effect	10 000	P: 99% s: TEG	Mayer (1987)

Mummichog

Fundulus heteroclitus	55 mm 1.7 g	24	st.; underground wellwater; 24°/oo; 20 °C; pH 8; DO 7.1-7.7	m, LC$_{50}$	> 85 100	P: act. ingredient s: acetone	Eisler (1970a)
Fundulus heteroclitus	55 mm 1.7 g	48	st.; underground wellwater; 24°/oo; 20 °C; pH 8; DO 7.1-7.7	m, LC$_{50}$	85 200	P: act. ingredient s: acetone	Eisler (1970a)
Fundulus heteroclitus	55 mm 1.7 g	96	st.; underground wellwater; 24°/oo; 20 °C; pH 8; DO 7.1-7.7	m, LC$_{50}$	58 000	P: act. ingredient s: acetone	Eisler (1970a)

Table 18 (continued)

Species	Life stage	Test period (h)	Experimental conditions	Criterion effect measured[a]	Concentration (μg/litre)	Remarks[b]	References
Fundulus heteroclitus	42 mm	96	st.; underground wellwater; 24°/oo; 20 °C; pH 8; DO 7.1-7.7	m, LC$_{50}$	8000	P: act. ingredient s: acetone	Eisler (1970b)
Fundulus heteroclitus	42 mm	96 (+240h observation)	st.; underground wellwater; 24°/oo; 20 °C; pH 8; DO 7.1-7.7	m, LC$_{50}$	4000	P: act. ingredient s: acetone	Eisler (1970b)
Fundulus heteroclitus	42 mm	96	st.; underground wellwater; 24°/oo; 20 °C; pH 8; DO 7.1-7.7	m, LC$_{50}$	1210	solution aged for 96 h	Eisler (1970b)
Fundulus heteroclitus	42 mm	96	10 °C	20% M	8000	P: act. ingredient s: acetone	Eisler (1970b)
Fundulus heteroclitus	42 mm	96	15 °C	10% M	8000	P: act. ingredient s: acetone	Eisler (1970b)

Table 18 (continued)

Species	Life stage	Test period (h)	Experimental conditions	Criterion effect measured[a]	Concentration (µg/litre)	Remarks[b]	References
Fundulus heteroclitus	42 mm	96	36 °/oo	100% M	8000	P: act. ingredient s: acetone	Eisler (1970b)
Striped killifish							
Fundulus majalis	84 mm 6.5 g	24	st.; underground wellwater; 24°/oo; 20 °C; pH 8; DO 7.1-7.7	m, LC₅₀	29 000	P: act. ingredient s: acetone	Eisler (1970a)
Fundulus majalis	84 mm 6.5 g	48	st.; underground wellwater; 24°/oo; 20 °C; pH 8; DO 7.1-7.7	m, LC₅₀	19 400	P: act. ingredient s: acetone	Eisler (1970a)
Fundulus majalis	84 mm 6.5 g	24	st.; underground wellwater; 24°/oo; 20 °C; pH 8; DO 7.1-7.7	m, LC₅₀	13 800	P: act. ingredient s: acetone	Eisler (1970a)

Table 18 (continued)

Fundulus heteroclitus	42 mm	96	20 °C	50% M	8000	P: act. ingredient s: acetone	Eisler (1970b)
Fundulus heteroclitus	42 mm	96	25 °C	100% M	8000	P: act. ingredient s: acetone	Eisler (1970b)
Fundulus heteroclitus	42 mm	96	30 °C	100% M	8000	P: act. ingredient s: acetone	Eisler (1970b)
Fundulus heteroclitus	42 mm	96	12 °/oo	0% M	8000	P: act. ingredient s: acetone	Eisler (1970b)
Fundulus heteroclitus	42 mm	96	18 °/oo	0% M	8000	P: act. ingredient s: acetone	Eisler (1970b)
Fundulus heteroclitus	42 mm	96	24 °/oo	10% M	8000	P: act. ingredient s: acetone	Eisler (1970b)
Fundulus heteroclitus	42 mm	96	30 °/oo	70% M	8000	P: act. ingredient s: acetone	Eisler (1970b)

Table 18 (continued)

Spot

Leiostomus xanthurus	84 mm 6.5 g	96	st.; natural seawater; 20°/oo; 25 °C; DO 3.2-4.5	m, LC$_{50}$	93	P: 99% s: TEG	Mayer (1987)
Leiostomus xanthurus	84 mm 6.5 g	96	st.; natural seawater; 20°/oo; 25 °C; DO 3.2-4.5	no effect	56	P: 99% s: TEG	Mayer (1987)
Leiostomus xanthurus	84 mm 6.5 g	96	flow-through 20°/oo; 25 °C	m, LC$_{50}$	59	P: 99% s: TEG	Mayer (1987)

Atlantic silverside

Menidia menidia	50 mm 0.8 g.	24	st.; underground wellwater; 24°/oo; 20 °C; pH 8; DO 7.1-7.7	m, LC$_{50}$	24 800	P: act. ingredient s: acetone	Eisler (1970a)
Menidia menidia	50 mm 0.8 g	48	st.; underground wellwater; 24°/oo; 20 °C; pH 8; DO 7.1-7.7	m, LC$_{50}$	21 900	P: act. ingredient s: acetone	Eisler (1970a)
Menidia menidia	50 mm 0.8 g	96	st.; underground wellwater; 24°/oo; 20 °C; pH 8; DO 7.1-7.7	m, LC$_{50}$	5700	P: act. ingredient s: acetone	Eisler (1970a)

Table 18 (continued)

Species	Life stage	Test period (h)	Experimental conditions	Criterion effect measured[a]	Concentration (µg/litre)	Remarks[b]	References
Striped bass							
Morone saxatilis	1 year	24	st.; 20 °C; pH 7.2; DO 6; H 50	m, LC₅₀	16 800	P: act. ingredient s: acetone	Rehwoldt et al. (1977)
Morone saxatilis	1 year	48	st.; 20 °C; pH 7.2; DO 6; H 50	m, LC₅₀	14 200	P: act. ingredient s: acetone	Rehwoldt et al. (1977)
Morone saxatilis	1 year	96	st.; 20 °C; pH 7.2; DO 6; H 50	m, LC₅₀	14 000	P: act. ingredient s: acetone	Rehwoldt et al. (1977)
Morone saxatilis	adult	96	interm. flow 12.8 °	m, LC₅₀	790	P: 99%	Earnest (1970)
Morone saxatilis	juvenile	96	flow-through 13 °C; 30 °/oo	m, LC₅₀	790	P: 86% s: ethanol	Korn & Earnest (1974)
Black mullet							
Mugil cephalus	48 mm 0.78 g	24	st.: underground wellwater; 24 °/oo 20 °C; pH 8; DO 7.1-7.7	m, LC₅₀	39 000	P: act. ingredient acetone	Eisler (1970e)

Table 18 (continued)

Mugil cephalus	48 mm 0.78 g	48	st.: underground wellwater; 24 °/oo 20 °C; pH 8; DO 7.1-7.7	m, LC_{50}	26 300	P: act. ingredient acetone	Eisler (1970a)
Mugil cephalus	48 mm 0.78 g	96	st.: underground wellwater; 24 °/oo 20 °C; pH 8; DO 7.1-7.7	m, LC_{50}	5200	P: act. ingredient acetone	Eisler (1970a)
Northern puffer						
Sphaeroides maculatus	196 mm 153 g	24	st.; underground wellwater; 24 °/oo 20 °C; pH 8; DO 7.1-7.7	m, LC_{50}	100 000	P: act. ingredient s: acetone	Eisler (1970a)
Sphaeroides maculatus	196 mm 153 g	48	st.; underground wellwater; 24 °/oo 20 °C; pH 8; DO 7.1-7.7	m, LC_{50}	91 000	P: act. ingredient s: acetone	Eisler (1970a)
Sphaeroides maculatus	196 mm 153 g	96	st.; underground wellwater; 24 °/oo 20 °C; pH 8; DO 7.1-7.7	m, LC_{50}	75 800	P: act. ingredient s: acetone	Eisler (1970a)

135

Table 18 (continued)

Species	Life stage	Test period (h)	Experimental conditions	Criterion effect measured[a]	Concentration (μg/litre)	Remarks[b]	References
Bluehead							
Thalassoma bifasciatum	90 mm 7 g	24	st.; underground wellwater; 24 °/oo 20 °C; pH 8; DO 7.1-7.7	m, LC$_{50}$	98 000	P. act. ingredient	Eisler (1970a)
Thalassoma bifasciatum	90 mm 7 g	48	st.; underground wellwater; 24 °/oo 20 °C; pH 8; DO 7.1-7.7	m, LC$_{50}$	88 000	P. act. ingredient	Eisler (1970a)
Thalassoma bifasciatum	90 mm 7 g	96	st.; underground wellwater; 24 °/oo 20 °C; pH 8; DO 7.1-7.7	m, LC$_{50}$	12 300	P. anct. ingredient s: acetone	Eisler (1970a)

Table 18 (continued)

AMPHIBIA

Rana cyano-phlyctis (male)	adult	96	st.; tap-water: 23 °C; pH 7.3-7.8; H 60-70; DO 6.7-7.9	m, LC_{50}	8000	Mudgall & Patil (1987)
Rana cyano-phlyctis (female)	adult	96	st.; tap-water: 23 °C; pH 7.3-7.8; H 60-70; DO 6.7-7.9	m, LC_{50}	11 500	Mudgall & Patil (1987)
Western chorus frog						
Bendacris triseriata	tadpole	96	st.; 15 °C; pH 7.1; H 44	m, LC_{50}	3700	Mayer & Ellersieck (1986)

[a] Criterion: m = mortality; i = immobilization; d = development; % M = % mortality.
[b] P = purity; s = solvent; TEG = triethylene glycol; techn.gr. = technical grade.
[c] st. = static.
[d] ‰ = salinity.
[e] H = hardness in mg/litre $CaCO_3$.
[f] DO = dissolved oxygen in mg O_2/litre.
[g] MATC = maximum acceptable toxic concentration.

Many laboratory studies have been performed on the acute toxicity of methyl parathion for fish. The following symptoms of methyl parathion poisoning can be expected to occur in fish: darkening of the skin, hyperactivity, body tremors, lethargy, jerky swimming, scalosis, loss of equilibrium, opercular or gaping paralysis, and death (Rao et al., 1967; Anees, 1975; Midwest Research Institute, 1975). One response that may be considered to be somewhat characteristic of acute methyl parathion poisoning in fish is the extreme forward position of the pectoral and pelvic fins (Midwest Research Institute, 1975; Srivastava & Singh, 1981).

Murty et al. (1984) state that the lowest concentration causing irreversible effects in the fish *Mystus carasius* after a 1-h exposure was 15 mg/litre.

7.2.1 Short-term toxicity in aquatic invertebrates

7.2.1.1 Laboratory studies on single species

Exposure of the freshwater mussel (*Lamellidens marginalis*) to sublethal (8 mg/litre) concentrations resulted in a transient increase (at 12 h) followed by a decrease (at 24-72 h) in the rate of respiration (Moorthy et al., 1984). Exposure of this species to concentrations ranging from 10 to 50 mg/litre resulted in a concentration-dependent decrease in heart rate (Rao et al., 1983a).

For crustaceans, long-term toxicity levels appear to be of the same magnitude as acute: a no-effect level on the reproduction of *Daphnia magna* was 0.0012 mg methyl parathion/litre after 21 days (artificial water, 18 °C; Dortland, 1980).

Exposure of the freshwater crab (*Oziotelphusa senex senex*) to sublethal levels of methyl parathion (0.1-1 mg/litre) resulted in complete inhibition of molt, a delay in the onset of molt, or a decrease in the percentage of molting animals (Reddy et al., 1985). A decrease in the carbohydrate content and increase in acid phosphatase activity in both the hepatopancreas and muscle also occurred (Reddy et al., 1986a; 1986b).

Eisler (1970a,b) found a 20% increase in mortality in *Nassa docoleta* after 10 days' exposure to 25 mg/litre (well water with a salinity of 24°/oo, 20°C, pH 8) in.

Exposure of prawns (*Penaeus indicus* or *Metapenaeus monoceros*) to sublethal concentrations of methyl parathion resulted in a concentration-dependent inhibition of acetylcholinesterase activity, which recovered in 7 days (Reddy & Rao, 1988). An increase in tissue levels of ammonia, urea, and glutamine, apparently resulted from the increased production of ammonia from purines and glutamate (Reddy et al., 1988; Reddy & Rao, 1990a). There was also an increase in tissue levels of fatty acids and cholesterol (Reddy & Rao, 1989), while the activity of alkaline phosphatase in the hepatopancreas was inhibited, and the acid phosphatase activity, enhanced (Reddy & Rao, 1990b). Changes in hepatic glycogen content and haemolymph glucose levels were observed after 5 days of sublethal methyl parathion exposure (Reddy & Rao, 1990b).

Cripe et al. (1981) tested the stamina of mysid shrimp (*Mysidopsis bahia*) in swimming against a water current in the presence of methyl parathion. Concentrations of 0.10 and 0.31 μg/litre did not affect maximum sustained speeds of the shrimp, but they were significantly reduced on exposure to 0.58 μg/litre.

2.1.2 Mesocosmic studies

After treatment of ponds with methyl parathion, the effects on daphnids were similar to those observed in the laboratory. However, indirect biological effects occurred that could not be predicted on the basis of laboratory tests. For example, the observed increase in populations of the crustacean *Diaptomus* sp. in treated ponds was attributed to the mortality of competitors (*Daphnia* spp.) and predators (*Cyclops* and aquatic insects) (Crossland & Elgar, 1983). Generally, recovery of zooplankton occurred soon after the end of treatment of ponds (Apperson et al., 1976; Crossland & Elgar, 1983). The numbers of free-swimming *Diptera* and *Ephemeroptera* were significantly reduced compared with controls, as were the benthic chironomid larvae in ponds treated at 100 μg/litre. Seventy days after treatment, there was evidence of recovery of populations of chironimids and *Ephemeroptera*, with full recovery 90 days after treatment (Crossland & Elgar, 1983).

7.2.2 Fish

7.2.2.1 Laboratory studies on single species

Jarvinen & Tanner (1982) conducted a long-term mortality study on the fish *Pimephales promelas* (flow through conditions, sterile water, 25 °C, pH 7.4-7.8, 46 mg $CaCO_3$/litre, 6.5-8.4 mg dissolved O_2/litre). Methyl parathion concentrations of 0.59-0.77 mg/litre induced increased mortality after 32 days. No effects on mortality were found at 0.38 mg/litre for the technical grade product and 0.59 mg/litre for the controlled release formulation. Mortality in rainbow trout *(Salmo gairdneri)* increased to 98% after exposure to 2.8 mg technical grade methyl parathion/litre (wellwater, 12 °C, pH 7.5, 272 mg $CaCO_3$/litre) for 96 h, followed by 7 days of observation (Palawski et al., 1983).

Exposure of the tilapia fish *(Tilapia mossambica)* to methyl parathion at a concentration of 0.09 mg/litre for 48 h resulted in a decrease in various anions and cations in tissues (Rao et al., 1983b), and in inhibition of acetylcholinesterase (20-60%) and ATPase (10-14%) activities. The activities of aspartate and alanine amino-transferase in muscle, gill, liver, and brain increased by 12-31% and 9-31%, respectively (Rao & Rao, 1984a; 1984b). Concentrations of carbohydrate and glycogen decreased in the tissues examined (Rao & Rao, 1983). Levels of soluble protein and the activity of glucose-6-phosphate dehydrogenase, a key enzyme of the hexose monophosphate shunt, in muscle, gill, and liver, were increased (Rao & Rao, 1987). Changes in carbohydrate metabolism were also observed in the freshwater fish *Clarias batrachus*, when exposed to sublethal concentrations of methyl parathion (7 mg/litre) for 48 and 96 h (Rani et al., 1989). There were significant decreases in glycogen (liver) and in pyruvate (liver, brain, gill) contents and increases in glucose (gill) and lactate (liver, brain, gill) levels, and the specific activities of several enzymes were inhibited.

Exposure of the catfish *(Channa punctatus)* to 52 μg methyl parathion/litre resulted in the elevation of serum triiodothyronine (T_3) as well as depression of brain acetylcholinesterase activity (Ghosh et al., 1989). This low dose of methyl parathion also impaired the regulation of gonadal function by gonadotropic hormone and gonado-tropin-releasing hormone in *Channa punctatus* (Ghosh et al., 1990). The inhibiting effect was also seen under field conditions where water concentrations of methyl parathion amounted to 0.239 μg/litre (Ghosh et al., 1990).

Exposure to sublethal doses of 0.1 mg methyl parathion/litre (corresponding to 1/5th of the LC_{50} values) for 75 days produced severe ovarian damage in the carp minnow (*Rasbora daniconius*) (Rastogi & Kulshrestha, 1990). Effects included diminished growth of ovaries and histopathological changes in immature, maturing, and mature oocytes.

Sublethal concentrations of methyl parathion (1.2 mg/litre) induced behavioural abnormalities in the juveniles of the fish *Cyprinus carpio*, such as imbalance, increased opercular movement and irritation (Babu et al., 1986). Exposed juveniles, when transferred to pesticide-free medium, showed rapid recovery.

Little et al. (1990) exposed rainbow trout (*Oncorhynchus mykiss*) to methyl parathion at 0.01 or 0.1 mg/litre and measured various behavioural parameters. Swimming capacity (as cm/s) was unaffected at any concentration tested, though spontaneous swimming activity was significantly reduced at both exposures. Number of prey (*Daphnia*) consumed was reduced, even at the lower exposure (0.01 mg/litre), but the percentage of daphnia consumed and the strike frequency of the fish on daphnia were only affected at 0.1 mg/litre. The capacity of the trout to escape from a predator was only reduced at 0.1 mg/litre.

In a static system (well water, salinity: 24°/oo, 20 °C, pH 8), with the fish *Fundulus heteroclitus*, the LC_{50} was 0.96 mg/litre after exposure for 10 days or 4 mg/litre after exposure for 4 days followed by 10 days in clean water (Eisler, 1970b).

2.2.2 *Mesocosmic studies*

In a methyl parathion-treated experimental pond, a high mortality rate was observed in rainbow trout, 37 days after treatment, which was associated with depression of the concentration of dissolved oxygen to less than 3 mg/litre, and decay of large amounts of algal biomass (Crossland & Elgar, 1983; see also section 7.3).

7.2.3 Amphibians

After application of methyl parathion to *Rana cyanophlyctis*, Mudgall & Patil (1987) found increased levels of glycogen in muscles, liver, and kidney, compared with control animals. On the basis of the marked elevated glycogen concentration in the kidney, it was concluded that the kidneys were the main target organ.

The effects of metacid (DDT + 50% w/w methyl parathion) on the development of the Indian bullfrog (*Rana tigrina*) were determined by Mohanty-Hejmadi & Dutta (1981). Threshold concentrations for adverse effects on eggs, feeding stage, and limb bud stage tadpoles ranged from 0.00005% to 0.004% metacid. These levels were much lower than the recommended dosage for the field application of metacid (0.15%).

7.3 Terrestrial organisms

7.3.1 Plants

Methyl parathion has been found to have phytotoxic effects in diverse crops, such as cotton (*Gossypium hirsutum*) (Brown et al., 1962; Roark et al., 1963; Youngman et al., 1989, 1990) and lettuce (*Lactuca sativa*) (Toscana et al., 1982; Johnson et al., 1983; Youngman et al., 1989).

Swamy & Veeresh (1987) found a reduction in lipid synthesis in methyl parathion-treated seeds of *Sorghum* sp., 24 h after germination. An increase in lipid production with a substantial elevation in unsaturated fatty acids was observed in methyl parathion-treated sorghum, 120 h after germination. The same effect occurred in 48-h seedlings, which were treated with the degradation products of methyl parathion. From this, it was concluded that the time-related reversal effect of methyl parathion is triggered by the pesticide degradation products themselves.

Exposure of sorghum seeds to methyl parathion for 1 h before germination resulted in an accumulation of proline in the seedlings and a reduction in growth, without affecting the water content. Residues of methyl parathion in the soil also influenced seed germination and seedling growth (Deshpande & Swamy, 1987).

7.3.2 Invertebrates

Poisoning of bees has been reported after incorrect application of methyl parathion on windy days (Bubien, 1971).

Analysis of dead honey bees (*Apis mellifera*; Hymenoptera) for pesticide residues, during 1983-85 in the USA, showed that the health of colonies, poisoned with methyl parathion (Penncap-M) or with a combination of methyl parathion and other insecticides, was often severely affected, whereas colonies contaminated by insecticides other than methyl parathion often recovered (Anderson & Wojtas, 1986).

Acute toxicity values were established for acetone formulations of methyl parathion applied topically to workers of Africanized and European honey bees (*Apis mellifera*) (Danka et al., 1986). The LC_{50} values of 0.32 μg and 0.17 μg/bee, respectively, showed the greater tolerance to methyl parathion of Africanized bees compared with European bees.

Jepson (1989) calculated a hazard ratio (ratio of contact LD_{50} at 0.11 μg/bee to the application rate of the pesticide at 500 g a.i./ha) for methyl parathion in honey bees of 8937 (using the method of Smart & Stevenson, 1982). Values of the hazard ratio greater than 50 are usually considered to indicate danger for bees. Along with azinphos methyl, methyl parathion has a very high indication of danger for bees from field spraying. Although the intrinsic toxicity for bees is as high for other pesticides, such as the pyrethroids, the hazard ratio is lower, since application rates of these pesticides are also lower.

Methyl parathion applied to small barriered plots of spring wheat at 1000 g a.i./ha did not have any apparent adverse effects on leaf litter decomposition and on earthworm populations (species not differentiated). Effects on individual earthworm species could not be demonstrated, because of statistically insufficient numbers of mature specimens collected (Shires, 1985).

Methyl parathion has adverse effects on many different beneficial insects. It was placed in the highest class of toxicity (score 4 in a classification of 1-4) for *Chrysopa* (Plannipennia), Coccinellidae (Coleoptera), and Hymenoptera (*Entomophaga*) (Höbaus, 1987). Side effects on the predator mite *Phytoseiulus persimilis* were placed in class 3 (Kniehase & Zoebelein, 1990).

Thompson & Gore (1972) assessed the toxicity of methyl parathion (95-99% purity) for *Folsomia candida* (Collembola) by direct contact in a spray tower and when applied to soil. In the direct-contact study, a 0.01% methyl parathion solution caused a 100% mortality of the collembola, 24 h after being treated. A 100% mortality rate also occurred in soil (Plainfield sand) treated with 0.5 mg methyl parathion/kg dry weight soil after a 24-h exposure.

Methyl parathion (0.05%) sprayed on coconut leaflets was found to be highly toxic to the parasitoid fauna (Hymenoptera; *Entomophaga*) of a coconut coccid (*Opisina arenosella*; Homoptera). The mortality of the caged insects was assessed 24 h after introduction of leaflets and after longer periods (Jalaluddin & Mohanasundaram, 1989).

Flanders et al. (1984) conducted a field study of methyl parathion (sprayed at recommended rates of 0.84 kg a.i./ha, in an encapsulated formulation, on soybeans) effects on *Pediobius foveolatus*, a parasitoid of the Mexican bean beetle *Epilachnia varivestis*. The pupae within parasitized beetles were unaffected by the insecticide and emerged normally. However, residues of methyl parathion on the plants killed 100% of the adult parasites emerging within 1 day, and 50% of those emerging within 3 days of the spraying. By 9 days after spraying, the mortality of emerging parasite adults was no longer affected by residues.

Walker et al. (1985) examined the effects of methyl parathion, used at 0.6 kg a.i./ha on rice fields in Louisiana, on the survival and reproduction of parasitic nematodes (*Romanomermis culcivorax*), introduced into the fields to control mosquito larvae. There were not any adverse effects of the insecticide on the nematodes.

Only a few cases of resistance to methyl parathion have been reported among arthropod parasites or predators. The reports refer to the braconid *Bracon mellitor* (Hymenoptera), a parasite of the boll weevil (*Anthonomous grandii*), which developed low levels of resistance after 5 or more generations of selection in the laboratory, and to field populations of the coccinellid *Coleomegilla maculata* (Coleoptera) taken from cotton fields, treated extensively with methyl parathion for 2 decades (Croft, 1977).

One week after application of methyl parathion (1000 g a.i./ha) on small barriered plots of spring wheat, the number of predatory beetles (mainly 4 species of Carabidae and 3 genera of Staphylinidae) fell to about 10% of that in the untreated control plot. Recovery occurred between 4 and 6 weeks after application, but a further fall in numbers of predatory beetles was observed 8-12 weeks after application (Shires, 1985). This second reduction was attributed to an indirect effect of the treatments, causing removal of the predators' food supply (mainly cereal aphids).

7.3.3 Birds

The acute lethal toxicities of methyl parathion for birds are compiled in Table 19.

Percutaneous administration of methyl parathion was more toxic for young mallard ducks (*Anas platyrhynchos*) than oral (dietary) administration (Hudson et al., 1979).

Studies on mallard ducks (*Anas platyrhynchos*) have shown that methyl parathion can affect the brood-rearing phase by increasing mortality and causing behavioural changes (Fairbrother et al., 1988). At least 40% of young ducklings exposed to sub-lethal oral doses of methyl parathion (4 mg/kg body weight) died within 40 min in outdoor enclosures. Several activities (swimming, preening, feeding) of mothers and ducklings were changed in treated broods. Ducks (*Anas platyrhynchos*; *A. discors*; *Aix sponsa*) nesting in agricultural fields aerially treated with methyl parathion (1.4 kg a.i./ha) had a higher average daily rate of duckling losses than those nesting in untreated fields (Brewer et al., 1988).

Spraying of methyl parathion at 1.4 kg a.i./ha did not significantly reduce the hatchability of starling (*Sturnus vulgaris*) eggs and the number of young fledglings per nest. However, collectively, the number of fledglings from the treated field was significantly lower than that from the control field (Robinson et al., 1988).

Buerger et al. (1991) dosed wild bobwhite quail (*Colinus virginianus*) with methyl parathion at 0, 2, 4, or 6 mg/kg body weight by oral intubation and then released them into the wild. The birds were monitored for 14 days by radio telemetry. Only the birds receiving 6 mg methyl parathion/kg body weight showed significantly reduced survival and this was the result of predation rather than overt toxicity. Activity was not affected by any treatment. Survivors did not show any inhibition of brain cholinesterase activity after 14 days, compared with controls.

Table 19. Acute lethal toxicities of methyl parathion for birds

Species	Age	Oral LD$_{50}$ (mg/kg body weight[a])	Dietary LC$_{50}$ mg/kg[b]	References
Mallard duck (Anas platyrhynchos)	5 days	8		Fairbrother et al. (1988)
	3 months	10		Hudson et al. (1984)
	adult	6.6		
Mallard duck (Anas platyrhynchos)	10 days		682	Hill et al. (1975)
Mallard duck (Anas platyrhynchos)	5 days		336	Hill et al. (1975)
Kestrel (Falco sparverius)	> 8 months	3.08		Rattner & Franson (1984)
Bobwhite quail (Colinus virginianus)	14 days		90	Hill et al. (1975)

Table 19 (continued)

Bobwhite quail (Colinus virginianus)	14 days	91	Bennet (1989)
Japanese quail (Coturnix coturnix japonica)	14 days	79	Hill et al. (1975)
Japanese quail (Coturnix coturnix japonica)	14 days	69	Hill & Camardese (1986)
Ring-necked pheasant (Phasianus colchicus)	10 days	91	Hill et al. (1975)
Red-winged blackbird (Agelaius phoeniceus)	-	10	Schafer (1972)

a intubation of a single dose.
b 8 days - standard test. 5 days feeding followed by 3 days observation.

Bennett et al. (1991) examined parameters of reproductive success in mallard ducks exposed to a dietary concentration of methyl parathion of 400 mg/kg. The female mallards were fed the methyl parathion diet at different stages of egg laying and incubation. Numbers of hatchlings per nest were 61%, 43%, and 58% of controls for birds exposed during egg laying, early incubation, and late incubation, respectively. Daily egg production was reduced during the treatment period, though 4 out of 10 hens resumed egg laying after treatment was terminated.

A dose-dependent inhibition of brain and plasma cholinesterase, hyperglycaemia, and elevated corticosterone concentrations were observed in the American kestrel (*Falco sparverius*) exposed to oral doses of up to 3 mg methyl parathion/kg body weight (Rattner & Franson, 1984).

Egg production in Japanese quail was inhibited and hatchability reduced at 60 mg/kg (NRC, 1977).

Methyl parathion-induced mortality following long-term ingestion was generally due to anorexia. Grackles (*Quiscalus quiscula*) had lost 28-36% of their initial body weight, when they died. No fat was visible and the muscles were reduced on the sternum. There was an increase in mortality at relatively constant intake rate of methyl parathion observed between May and August, which was related to an increase in natural activity within this time. It was concluded, that median lethal dietary concentrations are relative and depend on the anorexic and physiological condition of wild birds (Grue, 1982). The mean brain AChE activity of grackles (*Quiscalus mexicanus*) was significantly inhibited more than that of white-winged doves (*Zenaida asiatica*) and that of mourning doves (*Zenaida macroura*) after applications of EPN (phenylphosphonothioic acid *O*-ethyl *O*-*p*-nitrophenyl ester) and methyl parathion (Custer & Mitchell, 1987).

Free-living, female red-winged blackbirds (*Agelaius phoeniceus*) were captured on their nests and given oral doses of 2.4 or 4.2 methyl parathion mg/kg body weight and released immediately after dosing. Although methyl parathion caused ataxia, lacrimation, and lethargy and significantly depressed cholinesterase activity (> 35%) at 4.2 mg/kg, there were no apparent adverse effects on incubation behaviour and nesting success (Meyers et al., 1990).

Depressed brain acetylcholinesterase activity was also observed in 2 bird species (red-winged blackbird, (*Agelaius phoeniceus*), and dickcissel (*Spiza americana*)) inhabiting wheat fields treated with methyl parathion (0.67 kg a.i./ha). Maximal inhibition occurred 5 days after pesticide application. Enzyme activity levels returned to near normal levels by the tenth day following application. Cholinesterase inhibition for dickcissels and red-winged blackbirds differed significantly (74% versus 40%), and these differences could not be explained by the diets of the 2 species, as they were similar (Niethammer & Baskett, 1983).

A subacute oral dose of 3.5 mg methyl parathion/kg per day resulted in inhibition of brain cholinesterase (average decrease of 36%) in nuthatches (*Sitta carolinensis*) after 3-7 days exposure (Herbert et al., 1989).

7.3.4 Non-laboratory mammmals

Two wild rodent species (*Sigmodon hispudus* and *Mus musculus*) were found to have a higher mortality rate and to recover more slowly from exposure to methyl parathion at oral doses of 14-80 mg/kg, compared with laboratory rodents (Roberts et al., 1988). Clark (1986) reported a greater tolerance to methyl parathion in little brown bats (*Myotis lucifugus*) compared with wild mice (*Mus musculus*): the 24-h oral LD_{50} value (372 mg/kg body weight) of methyl parathion for little brown bats was 8.5 times the LD_{50} value for mice (44 mg/kg body weight). A loss of coordination was observed in 50% of the animals that were still alive 24 h after the treatment. The poisoned bats could be more easily captured by predators. The threshold of the coordination loss was about 1/3 of the LD_{50} value. In toxicity tests, mink (*Mustela vison*) rejected methyl parathion-treated diets and appeared to die from starvation rather than from methyl parathion poisoning (Aulerich et al., 1987).

8. EFFECTS ON EXPERIMENTAL ANIMALS AND *IN VITRO* TEST SYSTEMS

The inhibition by methyl parathion of acetylcholinesterase at nerve endings results in an accumulation of endogenous acetylcholine, as evidenced by peripheral and central cholinergic nervous system signs (Taylor, 1980).

Toxic effects include profuse salivation, lacrimation, nasal discharge, colic, diarrhoea, pupil constriction, excessive sweating, coughing, vomiting, frequent urination, anxiety, restlessness, hyperactivity, and hyperkinesis.

A more complete treatise on the effects of organophosphorus insecticides in general, especially their short- and long-term effects on the nervous system, can be found in *Environmental Health Criteria 63: Organophosphorus insecticides - A general introduction* (WHO, 1986).

8.1 Single exposure

Toxicological data on methyl parathion are summarized by Taylor (1980) and Flucke (1984).

The acute toxicity values in a number of species following the oral (Table 20), dermal (Table 21), inhalational (Table 22), and intraperitoneal (Table 23) administration of methyl parathion show lethal doses of about 3-400 mg/kg for the oral route, 40-300 mg/kg for the dermal route, 3.5-72 mg/kg for the intraperitoneal route, and 30-300 mg/m^3 for inhalation exposure. The acute subcutaneous LD$_{50}$s for methyl parathion in rats and mice were 6 and 18 mg/kg body weight, respectively (Krueger & Casida, 1957; RTECS, 1991); the acute intravenous LD$_{50}$ was reported to be 4.1-14.5 mg/kg body weight in rats, 2.3-13 mg/kg body weight in mice, and 50 mg/kg body weight in guinea-pigs (NIOSH, 1976).

Izmirova et al. (1984) found an abrupt reduction in the blood and brain cholinesterase and acetylcholinesterase activities in albino rats in the 30th and 90th min after a single oral administration of 32 mg methyl parathion/kg. The blood cholinesterase activity was reduced by 71% and the brain acetylcholinesterase activity by 54%. Twenty-four hours after administration, the cholinesterase activity was higher than that in the controls.

Table 20. Acute oral toxicity

Animal (sex)[a]	LD$_{50}$ (mg/kg body weight)	References
rat (m) fasted	2.9	Heimann (1982)
rat (f) fasted	3.2	Heimann (1982)
rat	6	Bayer AG (1988); RTECS (1991)
rat (m)	7.4	Flucke & Kimmerle (1977)
rat (f) nonfasted	9.3	Heimann (1982)
rat (m) nonfasted	10.8	Heimann (1982)
rat (m)	11.7	Kimmerle (1975)
rat (m)	14	Gaines (1960, 1969)
rat (f)	24	Gaines (1960, 1969)
rat	35	Kagan (1971)
mouse	23	RTECS (1991)
mouse	14.5	Haley et al. (1975)
mouse	33.1	Mundy et al. (1978)
mouse	21.8	Mundy et al. (1978)
mouse	19.5	Haley et al. (1975)
rabbit (m) fasted	19	Heimann (1982)
rabbit (f) fasted	19.4	Heimann (1982)
rabbit	420	RTECS (1991)
guinea-pig	1270	RTECS (1991)
guinea-pig	417	NIOSH (1976)
dog	90	Hirschelmann & Bekemeier (1975)

[a] m = male, f = female.

Table 21. Acute dermal toxicity

Animal, sex[a]	Duration of exposure (h)	LD_{50} (mg/kg body weight)	LD_{100}	References
rat	1[b]	63		RTECS (1991)
rat (m,f)	-	67		Gaines (1960, 1969)
rat (m)	24	46		Heimann (1982)
rat (f)	24	44		Heimann (1982)
rabbit (m)	6		1270 (pure)	Deichmann et al. (1952)
rabbit (m)	6		350-780 (technical grade)	Deichmann et al. (1952)
rabbit (m)	6		420 (pure, in corn oil)	Deichmann et al. (1952)
rabbit (m)	6		2500 (pure, suspended in water)	Deichmann et al. (1952)
rabbit	-	300		RTECS (1991)

[a] m = male, f = female.
[b] no data given.

Table 22. Acute inhalation toxicity

Animal (sex)[a]	Duration of exposure (h)	LC_{50} (mg/m^3 air)	References
rat	1	120	RTECS (1991)
rat	1	34	Molnar & Paksy (1978)
rat (m)	1	200	Kimmerle & Lorke (1968)
rat (m)	1	260	Thyssen (1979)
rat (f)	1	320	Thyssen (1979)
rat (m)	4	120	Kimmerle & Lorke (1968)
rat (m)	4	185	Thyssen (1979)
rat (f)	4	170	Thyssen (1979)
mouse	4	120	RTECS (1991)

[a] m = male   f = female.

Table 23. Acute intraperitoneal toxicity

Animal	LD_{50}(mg/kg body weight)	References
rat	3.5	Du Bois & Coon (1952)
rat adult	5.8	Brodeur & Du Bois(1963)
rat juvenile	3.5	Brodeur & Du Bois (1963)
rat	7	Kimmerle (1975)
mouse	9.3	Kimmerle (1975)
mouse	11.0	Benke et al. (1974)
mouse	6.4	Kamienski & Murphy (1971)
mouse	8.2	Mirer et al. (1977)
mouse	72	Goyer & Cheymol (1967)

Dogs that received 10 or 30 mg methyl parathion/kg body weight intravenously showed minimal activity of the plasma cholinesterases, 30 min after treatment. Sixteen hours after the injection of 10 mg methyl parathion/kg body weight, the enzyme activities had returned to their pre-injection values. However, following treatment with 30 mg methyl parathion/kg body weight, it took 7 days for complete recovery (Braeckman et al., 1980).

After i.p. injection of 2.4 mg Wofatox (methyl parathion), Karcsu et al. (1981) observed complete inhibition of the histochemically detectable acetylcholinesterase activity in the central nervous system of the rat. Partial enzyme inhibition was found in the motor neurons and in the striated muscles. Ultrastructural changes in the myocardium of the rats were also confirmed.

8.2 Skin and eye irritation, sensitization

The skin of rabbits exposed to methyl parathion for 4 or 6 h did not show perceptible signs of irritation (concentrations up to LD_{100}, Table 21, Deichmann et al., 1952). Similar results were obtained by Hecht & Wirth (1950) and by Heimann (1982) in their studies on rats.

The irritation potential of methyl parathion on the rabbit skin and eye was studied according the OECD guidelines for the testing of chemicals (Nos. 404 and 405). It was concluded that methyl parathion had no primary irritating potential (Pauluhn, 1983).

8.3 Short-term exposures

Groups of Wistar albino rats were exposed to methyl parathion aerosol concentrations of 0.9, 2.6, and 9.7 mg/m^3 air for 6 h/day, 5 days/week for 3 consecutive weeks. No mortality occurred. Plasma and brain cholinesterase levels were significantly depressed in the highest dose group. At 2.6 mg/m^3, slight inhibition of plasma ChE occurred (Thyssen & Mohr, 1982).

Groups of New Zealand white rabbits were administered methyl parathion (purity 96.3%) dermally at dose levels of 10, 50, and 250 mg/kg body weight, applied for 5 days/week over 3 weeks. The site was left uncovered for 6 h and then it was cleaned with soap and water. There was a dose-related inhibition of erythrocyte and brain cholinesterases in the 50 and 250 mg/kg dose groups. Plasma ChE was also significantly depressed in the highest dose group; these animals presented signs of cholinergic poisoning and 5 out of 6 animals died (Mihail & Vogel, 1984).

A 12-week dietary study at 5, 20, and 50 mg methyl parathion/kg was performed on male and female dogs. The doses corresponded to 0.1, 0.4, and 1.0 mg/kg body weight per day. A significant decrease in plasma cholinesterase activity was observed only at 50 mg/kg diet (Williams et al., 1959).

8.4 Long-term exposures

Kazakova et al. (1974) fed chicken and cattle daily with 2.5 mg methyl parathion/kg body weight for one year. No changes in health status and food intake were observed. In pigs and cows, 10 mg/kg body weight led to irritation, depression, miosis, salivation, intensified peristaltics, and diarrhoea.

Rats fed diets containing 40 mg methyl parathion/kg for 2 years, and mice (females: fed up to 125 mg/kg, males: fed up to 77 mg/kg) did not display any cholinergic toxicity (NCI, 1979).

In a 2- year study (combined long-term/carcinogenicity), 500 rats (50 male, 50 female per dose, 100 controls) were fed diets containing 0, 2, 10, or 50 mg methyl parathion/kg. The intake of active ingredient was 0, 0.144, 0.713, 4.917 mg/kg body weight per day (females). The highest dose led to retardation of growth, increase in mortality, inhibition of cholinesterase-activity in plasma, erythrocytes, and brain, reduction of haemoglobin, and haematocrit, and an increase in reticulocytes, after 2 years. Female rats showed a reduction in plasma proteins and a reversible increase in urea in plasma and protein in urine. At 10 mg/kg diet, the cholinesterase activity in plasma and red blood cells was inhibited. Male rats also showed reduced cholinesterase activity in the brain. Extensive histopathological examinations (cardiovascular, respiratory, and urogenital systems, digestive tract, organs, and glands) did not exhibit any substance-related changes. No toxic effects were found at the lowest dose (Bomhard et al., 1981; Schilde & Bomhard, 1984).

Sixty rats per sex and group were fed diets containing methyl parathion at concentrations of 0.5, 5, or 50 mg/kg for 2 years. Sciatic nerve preparations from 1 out of 5 males in the low-dose group and 1 out of 5 in the mid-dose group reportedly showed moderate degenerative changes. In the high-dose group (50 mg/kg diet), sciatic nerve preparations from treated males showed a loss of myelinated fibres. These animals also showed more myelin degeneration and Schwann cell proliferation. Similar, less severe changes were seen with a lower incidence, in males fed 5.0 or 0.5 mg/kg per day males and in the controls. Only 1 rat in the low-dose group and 1 in the mid-dose group had more severe changes than the controls; however, 4 high-dose males showed more severe changes. No obvious differences were seen in the females. Haemoglobin, haematocrit, and RBCs were slightly reduced in mid-and high-dose males, and moderately reduced in high-dose females (Daly, 1983).

8.5 Reproduction, embryotoxicity, and teratogenicity

Dosages of 4 or 6 mg methyl parathion/kg body weight were injected intraperitoneally into pregnant, female albino Holzmann rats. The injection was made on day 9 or day 15 using an ethanol-propylene glycol vehicle. It was found, that the fetal, cerebral, cortical cholinesterase activity was reduced, indicating the transplacental passage of the organophosphate. Large subcutaneous haematomas also occurred; however, no significant developmental defects were noticed (Fish, 1966). Ackermann (1974) also reported that there was no placental barrier for methyl parathion.

A 3-generation study was performed by the Woodard Research Corporation in 1966. This unpublished report was reviewed by Anon., FAO/WHO (1969). Rats received diets of 0, 10, or 30 mg methyl parathion per kg diet. A sporadic reduction in the litter size of groups was observed (30 mg/kg: $F_{2\alpha}$, $F_{2\beta}$, $F_{3\alpha}$; 10 mg/kg: $F_{1\beta}$), also a delayed growth of litters until weaning (30 mg/kg: $F_{2\alpha}$, $F_{3\alpha}$, $F_{3\beta}$; 10 mg/kg: $F_{1\beta}$), a reduced rate of survival of the litters (30 mg/kg: $F_{1\alpha}$, $F_{1\beta}$, $F_{2\alpha}$); 10 mg/kg: $F_{3\alpha}$), and an increased rate of stillbirths (30 mg/kg: $F_{1\beta}$, $F_{3\alpha}$).

Another 3-generation study was performed by the Midwest Research Institute (1975). Rats received 0, 10, or 30 mg methyl parathion/kg diet (corresponding to 0, 0.5, or 1.5 mg methyl parathion/kg body weight); 2 litters of each generation were evaluated. No adverse effects on growth, survival, or reproduction were observed at the 30 mg/kg level, however, the 10 mg/kg level caused a reduction in the postnatal survival in weaning rats in the $F_{1\beta}$ and $F_{3\alpha}$ generations. Similar results were found in the 3-generation study of Löser & Eiben (1982). Rats (male and female, SPF-Wistar W 74 strain, 5-6 weeks old) were fed a diet containing technical methyl parathion (95% pure) at 2, 10, or 50 mg/kg for 77 days and then mated. The no-effect level in this study was 2 mg/kg diet. A dose of 50 mg methyl parathion/kg caused reductions in neonatal weights and litter size, and delayed body weight gains, while 10 mg/kg caused sporadic reductions in litter size ($F_{2\alpha}$, $F_{3\alpha}$), delayed growth of litters until weaning ($F_{1\alpha}$, $F_{2\alpha}$, $F_{2\beta}$), and a reduced rate of survival of the litters.

Single doses of 3, 30, or 100 μg methyl parathion (in 10% DMSO) were administered subgerminally into chicken eggs on day 2 and intra-amniotically on days 3 and 4. These doses did not induce any specific malformations. Embryotoxicity was noted at the 2 higher doses (30 and 100 μg) (Benes & Jelinek, 1979). These findings were confirmed by estimating the embryotoxicity range and parameters (Jelinek et al., 1985). Doses of up to 55 μlitre Wofatox 50EC/kg egg reduced haematocrit, glucose, cholesterol, and AChE activity and increased aspartate aminotransferase and lactate dehydrogenase values in blood samples of chicken embryos (Somlyay et al., 1989). The injection of 2 different concentrations of methyl parathion (13 and 135 mg/kg egg) into pheasant eggs resulted in increased mortality and in an increased incidence of skeletal deformities in the survivors (Varnagy et al., 1984; Déli & Varnagy, 1985; Varnagy & Déli, 1985). Biochemical studies on muscle samples from chicken embryos (eggs treated with 0.4% or 4.0% solution of Wofatox 50EC) showed decreased creatine kinase activity, decreased creatine, creatine-phosphate, and Mg^{2+} (in cervical muscle only) contents, and increased creatinine, Ca^{2+}, and Mg^{2+} (in femoral muscle only) values (Déli et al., 1985). Scanning electron microscopic examination of the cartilage in chicken embryos showed degeneration of collagen structure and chondrocytes at a high insecticide concentration (eggs treated with 0.4% or 4.0% solution of Wofatox 50EC) (Varnagy et al., 1988). Analysis of the protein pattern of the cervical muscles of 18-day-old embryos, treated with 0.4% methyl parathion solution showed decreases in α-actinin, α-tubulin, β-tubulin, and γ-proteins (Déli & Kiss, 1988).

Studies on chickens showed that methyl parathion at 1-10 μmol/litre had no or only little effect on the adenylate cyclase in the embryo muscle. Comparable results were obtained with rats using the plasma membrane adenylate cyclase in rat livers, even at

100 μmol/litre. In the presence of adenylate cyclase-stimulating agents, additional activation of methyl parathion was observed; it enhanced the stimulating activity of GTP and isoproterenol together, but not alone. Methyl parathion is soluble, but not metabolized, in plasma membranes, so it may alter cellular levels of cAMP, and, thus, cell growth (Déli & Kiss, 1986).

At very high doses (20 or 60 mg/kg body weight), methyl parathion injected intraperitoneally in ICR-CL mice on day 10 of pregnancy, caused convulsions, hypersalivation, ataxia, and tremor. At the higher dose, 5 out of the 14 litters died. This dose caused reduced neonatal weight, an increase in the occurrence of cleft palate, and an increased incidence of cervical ribs in the fetuses. At the lower dose, cleft palate, and a statistically non-significant increase in the number of cervical ribs and underdeveloped sternebrae were observed (Tanimura et al., 1967).

A single intraperitoneal injection of 5, 10, or 15 mg/kg body weight was administered to Wistar rats on day 12 of pregnancy; signs of toxicity and reduced body weight were observed with 15 mg/kg, but there was no evidence of teratogenicity at any of the doses (Tanimura et al., 1967).

On 6 alternate days between days 5 and 15 of pregnancy, 3 groups of rats received orally 0.1, 1, or 3 mg methyl parathion/kg body weight. Another group of rats received 3 mg methyl parathion/kg body weight on 8 alternate days between days 5 and 19 of pregnancy. No teratogenic effects were observed; however, the high doses caused increased resorptions and decreased fetal body weight (Fuchs et al., 1976).

Methyl parathion was administered orally by gavage, to groups of female rats from day 6 to day 15 of gestation at dose levels of 0.1, 0.3, or 1 mg/kg body weight. Weight gain in the mothers and a slight retardation in growth in the fetuses were noted at the highest dose level. Methyl parathion was not toxic for the embryo or fetus and no teratogenic effects were apparent (Machemer, 1977a).

Groups of 24-26 rats received intravenous injections of 0, 0.03, 0.1, or 0.3 mg methyl parathion/kg body weight per day from day 6 to day 15 of pregnancy. On day 20, the fetuses were evaluated. No treatment-related effects were found (Machemer, 1977b).

No signs of embryotoxicity or teratogenicity were found in rabbits that received 0.3, 1.0, or 3.0 mg methyl parathion/kg body weight on days 6-18 of pregnancy (Renhof, 1984).

Daily intraperitoneal administration of methyl parathion (1 or 1.5 mg/kg body weight) to rats during days 6-19 of gestation resulted in decreases in both maternal and fetal protein synthesis (Gupta et al., 1984). The effect was dose dependent, and was greater on day 19 than on day 15 of gestation; it was also greater in fetal than in maternal tissues. The same dosage regimen resulted in a postnatal decrease in acetylcholinesterase activity and muscarinic receptor binding. Recovery of acetylcholinesterase activity to near normal levels occurred by day 28 in the low-dose offspring, but not in the high-dose weanlings (Gupta et al., 1985).

8.6 Mutagenicity and related end-points

Methyl parathion has been reported to have DNA-alkylating properties. Mutagenicity test results have been both positive and negative. The results of most of the *in vitro* mutagenicity studies with both bacterial and mammalian cells were positive; the *in vivo* studies produced equivocal results. A survey is given in Table 24.

8.7 Carcinogenicity

The carcinogenicity of methyl parathion was studied in mice by the National Cancer Institute (NCI) in 1979. Groups of 50, six-week-old female B6C3F$_1$ mice received diets containing either 62.5 or 125 mg methyl parathion/kg for 102 weeks. For 37 weeks, 2 groups of 50 male mice received diets containing either 62.5 or 125 mg methyl parathion/kg, which was reduced then to 20 or 50 mg/kg for another 65 weeks. Untreated matched groups of 20 males and 20 females were used as a control. From all groups, 80-86% were still alive at the end of the study. There was no statistically significant increase in tumour incidence.

The NCI (1979) also studied the carcinogenicity of methyl parathion in rats. Groups of 50 female and male Fischer 344 rats (6 weeks old) received separate diets containing 20 or 40 mg methyl parathion/kg for 105 weeks. As matched controls, 20 male and 20 female rats remained untreated. Only 46% of the high-dose

Table 24. Mutagenicity tests of methyl parathion

Test	Species	Dose levels	Metabolic activation	Results	Reference
	Microorganism				
Gene mutation tests Ames	S. typhimurium TA100, TA1535 TA1536, TA1537 TA1538	-ᵃ	+/- +/-	- -	Simmon et al. (1977) Carrere et al. (1978)
Ames	S. typhimurium TA100	250-1250 µg/plate	+	+	Rashid & Mumma (1984)
Ames	S. typhimurium TA98, TA1535, TA1537, TA1538	250-1250 µg/plate	+	-	Rashid & Mumma (1984)
Ames	S. typhimurium TA1535 TA100 TA1537, TA1538	≥ 1000 µg/plate ≥ 500 µg/plate 20-2500 µg/plate	+/- +/- +/-	+ + -	Herbold (1986)
Reverse mutations	E. coli WP2 and WP2uvrA	1 crystal or microdrop 250-2500 µg/plate	- +/-	- -	Dean (1972) Simmon et al. (1977) Rashid & Mumma (1984)
Forward mutations streptomycin/5-methyl-tryptophane resistance	E. coli	1 × 10⁻²mol/litre	-	+	Mohn (1973) Wild (1975)

Table 24 (continued)

ade-6 forward mutation	Sacharomyces pombe	11-228 mmol/litre		-	Gilot-Delhalle et al. (1983)
Recombinogenic activity/point mutation (8-Aza-guanine resistance)	Aspergillus nidulans	2 mg		-	Morpugo et al. (1977)
	Streptomyces coelicolor	-ᵃ		-	Carere et al. (1978)
Insects					
Sex link recessive lethal test	Drosophila melanogaster Larvae 24,48,72 h old	1.25×10^{-5} % w/w 6.3×10^{-6} % w/w 24-48 and 72 h exposure		+	Tripathy et al. (1987)
Mammals					
Thymidine kinase locus	Mouse lymphoma cells L51784	-ᵃ	-/+		Jones et al. (1982)
DNA effects chromatid exchange	Chinese hamster ovary cells V79 (in vitro)	20 and 40 µg/ml 28-72 h	-ᵃ	+	Chen et al. (1981)

161

Table 24 (continued)

Test	Species	Dose levels	Metabolic activation	Results	Reference
	Human lymphoid cells (LAZ-007)	20 μg/ml	-	+	Sobti et al. (1982)
	Human lymphoid cells B35 M and Jeff cells	20 and 40 μg/ml 28-72 h	-[a]	+ (for 20 μg/ml)	Chen et al. (1981)
Sister chromatid exchange (SCE)	Human lymphocytes	36-181.8 μmol/litre	-	+ (dose dependent)	Singh et al. (1987)
Unscheduled DNA synthesis	Human fetal lung WI 38 fibroblast	-[a]	+/-	-	Simmon et al. (1977)
	Human lymphoid cells B411-4 RMPI - 1788 RMPI - 7191	up to 50 μg 6-50 h	-	-	Huang (1973)
	Mouse (in vivo) bone marrow germ cells	10 mg/kg ip		-	Degraeve & Moutschen (1984)
	Mouse, Swiss bone marrow	9.4, 18.8, 37.5, 75 mg/kg body weight, oral		+ (dose dependent)	Mathew et al. (1990)

Table 24 (continued)

Chromosomal aberrations	Rat bone marrow cells (*in vivo*)	0.5 mg/kg body weight 1 mg/kg 2 mg/kg 5 days/week for 7 weeks	+ (dose 1.95% 9.26% 16.86% dependent)	Malhi & Grover (1987)
Micronucleus	Wistar rat (*in vivo*)	1, 2 and 4 mg/kg ip	+ (dose dependent)	Grover & Malhi (1985)
Micronucleus	Mouse	5-10 mg/kg body weight, orally, daily, for 2 days	-	Herbold (1986)
Micronucleus	Mouse, Swiss	9.4, 18.8, 37.5, 75 mg/kg body weight, orally	+ (dose dependent)	Mathew et al. (1990)
Dominant lethal	Mouse, male ICR/SIM	20 mg/kg diet 40 mg/kg for 80 mg/kg 7 weeks	-	Simmon et al. (1977)
Dominant lethal	Mice, male Q strain	0.15 mg/litre (daily) in drinking-water, 5-7 weeks	-	Degraeve et al. (1984)

[a] = no information given.

females survived, but 78% high-dose males, 74% low-dose males, 82% low-dose females, 85% control males, and 95% control females were still alive at the end of the study. No statistically significant increase in tumour rates was found.

Male and female rats in groups of 50 were fed for 2 years with diets containing 2, 10, or 50 mg methyl parathion/kg. No toxic effects were found at the low dose (see section 8.4). No morphological changes due to the insecticide were detected. No carcinogenic effects of methyl parathion were observed (Bomhard et al., 1981; Schilde & Bomhard, 1984).

8.8 Special studies

Seven white New Zealand rabbits per group were fed 0, 0.036, 0.162, 0.519, or 1.479 mg methyl parathion/kg body weight per day, for 8 weeks. A dose-dependent increasing atrophy of the thymus cortex and a reduced, delayed-type hypersensitivity response (DTH) were found (Street & Sharma, 1975). In a preliminary study, Fan et al. (1981) examined the effects of methyl parathion on immunological responses to *S. typhimurium* infection in mice. Mortality rates among infected animals fed 0.08, 0.3, or 0.7 mg methyl parathion/kg body weight (duration "extending beyond 2 weeks") were determined and protection by vaccination was examined. Dose-related increases in mortality were seen in unvaccinated mice and protection by immunization was decreased. These limited findings were reviewed by Sharma & Reddy (1987) and Thomas & House (1989).

Methyl parathion was found by Barnes & Denz (1953) not to cause delayed neuropathy in their hen test. However, Nagymajitenyi et al. (1988) found neurotoxic effects on the central and peripheral nervous systems in both acute and short-term studies on CFY rats, in which the conduction velocity of the peripheral nerves, muscle function (ischidiacus nerve/gastrocnemius muscle), and EEG activity were measured. In the short-term studies, the rats were given 0.44 mg methyl parathion/kg body weight for 5 days/week for 6 weeks; in the acute study, the rats received 0.4 mg/kg, orally.

Lipid metabolism in rats was investigated by Hasan & Ahmad Khan, (1985). The rats received daily intraperitoneal doses of 1.0, 1.5, or 2.0 mg methyl parathion/kg body weight for 7 days. The concentrations of total lipids, phospholipids, and cholesterol increased in a dose-related manner in the cerebral hemisphere, cerebellum, brain stem, and spinal cord. Lipid peroxidation increased in the CNS with the exception of the cerebellum.

Khan & Hasan (1988) studied changes in the levels of gangliosides and glycogen of the cerebral hemisphere, cerebellum, brain stem, and spinal cord following intraperitoneal injection of methyl parathion (1, 1.5, or 2 mg/kg body weight) in 24 rats for 7 days. A dose-related depletion in the concentration of gangliosides and glycogen content were discernible in all regions of the CNS.

Preweanling, male, rat pups were exposed daily through subcutaneous injection to parathion (1.3 or 1.9 mg/kg body weight) or the vehicle (corn oil) on postnatal days 5-20, a period critical for the development of behavioural and biochemical parameters of the cholinergic nervous system. This exposure resulted in dose-dependent reductions in acetylcholinesterase activity and muscarinic receptor binding in the cortex. During the preweanling period, there were no differences among the groups in most reflex measures, eye opening, or incisor eruption. Postweanling behavioural assessment revealed small deficits in tests of spatial memory in both the T-maze and radial arm maze. There were no differences in neuromuscular abilities or spontaneous activity measures (Stamper et al., 1988).

The behavioural effects of short-term exposure of male Wistar rats to methyl parathion (1/50 or 1/100 of LD_{50}, orally, for 6 weeks) were studied. Open-field (OF) and elevated plus-maze (EPM) tasks were used to decide whether or not the compound could affect behaviour. Significant effects were measured in OF activity during the first minute, on the activity of crossing outer squares, increasing latencies to leave centre, start of rearing, grooming, and defecation. EPM parameters showed an increased amount of time spent in the open arms and a clear tendency to enter open arms more frequently. The defecation rate in the EPM was significantly decreased (Schultz et al., 1990).

8.9 Factors modifying toxicity

Methyl parathion becomes toxic only after metabolic transformation to the oxon analogue, methyl paraoxon, by liver microsomal oxidation. The microsomal enzymes metabolize methyl parathion in 2 ways *in vitro*: *a*) oxidation to methyl paraoxon, and *b*) degradation to dimethyl phosphorothiotic acid and *p*-nitrophenol.

NADPH and O_2 are necessary for both reactions, indicating that these are oxidative processes (Nakatsugawa et al., 1968).

Piperonyl butoxide inhibits the mixed function oxidase activity of the microsomal fraction of liver cells. Therefore, it inhibits both oxidative activation of methyl parathion and detoxification, but not the dealkylation reactions due to glutathione-*S*-alkyltransferase. At a dosage of 400 mg/kg body weight, piperonyl butoxide antagonized the toxic effects of methyl parathion in mice when given 1 h before the mice received the insecticide. The intraperitoneal toxicity of methyl parathion was reduced 40-fold (Kamienski & Murphy, 1971; Levine & Murphy, 1977a,b; Mirer et al., 1977). Diethyl maleate reduced the glutathione content of the liver by 80%. This agent increased the acute toxicity of methyl parathion by the inhibition of glutathione-dependent detoxification (Mirer et al., 1977).

Pap et al. (1976) showed that methyl parathion was less toxic in rats with a thioacetamide-induced liver cirrhosis than in normal rats. After activating the microsomal enzymes in the liver with sodium phenobarbital or norandrostenolone phenylpropionate, the cirrhotic rats showed a normal susceptibility to methyl parathion, indicating the involvement of liver microsomes in the activation of methyl parathion. Treatment of normal rats with chloramphenicol could increase their survival time after poisoning with methyl parathion.

Lead nitrate ($Pb(NO_3)_2$) reduced the toxicity of methyl parathion due to an increase in the carboxylesterase-dependent metabolism of the insecticide (Hapke et al., 1978).

A single oral dose of 5 or 10 mg methyl parathion/kg body weight resulted in decreases in the cholinesterase activity in rats of 43.6% or 72.3%, respectively. However, rats pretreated on 5 successive days with a combination of 7 mg gentamycin/kg body weight and 20 mg rifamycin/kg body weight showed a remarkable protection against the toxic effects of methyl parathion. The toxic signs were minimal; the rats showed no, or only transient, signs of poisoning, and no convulsions were be observed in the rats that had been pretreated. The combination of these 2 drugs significantly prevented the methyl parathion-induced inhibition of cholinesterase in plasma and of the liver carboxylesterase. Gentamycin or rifamycin alone did not have any effect. Youssef et al. (1987) demonstrated, that gentamicin and rifamycin inhibited the formation of the oxidation product of methyl parathion, methyl paraoxon, in the liver and skeletal muscle. Both substances potentiated the rate of urinary *p*-nitrophenol excretion within 48 h of the methyl parathion application. Pretreatment with rifamycin influenced the rate of liver glutathion reduction, whereas gentamicin did not show this effect.

Male rats were treated with a single i.p. dose of 5 mg methyl parathion/kg. Pretreatment with memantine hydrochloride (18 mg/kg, i.p.), 30 min before methyl parathion administration, and atropine sulfate (16 mg/kg, i.p.), 15 min before, significantly reduced ($P < 0.01$) the inhibition of acetylcholinesterase (Gupta & Kadel, 1990).

Pretreatment with cimetidine, which suppresses the hepatic microsomal oxidative metabolism, decreased the toxicity of methyl parathion in rats and mice (Joshi & Thornburg, 1986).

Fuchs et al. (1986) showed that the LD_{50} in rats increased by 19-24% after simultaneous oral administration of 0.5 g humic acids/kg body weight and methyl parathion. It was supposed that the absorption of methyl parathion from the digestive tract decreased as a result of the intake of the humic acids.

Sultatos (1987) perfused mouse livers *in situ* with methyl parathion. The acute toxicity of methyl parathion in mice was antagonized by pretreatment with phenobarbital, daily, for 4 days (80 mg/kg, i.p.). This effect was due to hepatic microsomal activation and resulted in an increased clearance of methyl parathion. Similar results were obtained by Du Bois & Kinoshita (1968), Du Bois (1969, 1971), and Murphy (1980).

The influence of temperature on the toxicity of methyl parathion in mice was studied by Nomiyama et al. (1980). They found median lethal doses (i.p.) of 14 mg/kg body weight at 8 °C, 44 mg/kg body weight at 22 °C, and 35 mg/kg body weight at 38 °C.

An influence of age on the toxicity and metabolism of methyl parathion was observed by Benke & Murphy (1975) in rats. Rats became much less sensitive to poisoning with methyl parathion with increasing age. The effect was explained by a presumable increase in the detoxification processes as the GSH-dependent (glutathion-dependent) dealkylation. Methyl parathion dealkylation rates increased directly with age for both sexes.

Carbon disulfide pretreatment, 1 h before administration of 10 mg methyl parathion/kg body weight to mice did not significantly affect the methyl parathion toxicity (Yasoshima & Masuda, 1986).

Prior depletion of glutathione by acetaminophen (600 mg/kg, i.p., Costa & Murphy, 1984) or by diethyl maleate (1 ml/kg, i.p., Sultatos & Woods, 1988) had little effect on the toxicity of methyl parathion (2.5 mg/kg body weight and up to 55 mg/kg body weight i.p., respectively) in the mouse.

Interactions of organophosphorus pesticides and several pyrethroid insecticides were reported by Gaughan et al. (1980). Following an intraperitoneal injection of organophosphorus pesticides in mice, they found pronounced inhibition of the liver microsomal esterase, which hydrolyses trans-permethrin. Methyl parathion did not potentiate the toxicity of deltamethrin (Audegond et al., 1989). Equitoxic oral or i.p. combinations of methyl parathion with other organophosphorus insecticides (amiton, coumaphos, crufomat, dimethoate, dioxathion, disulfoton, fensulfothion, ethyl parathion, phosphamidon, trichlofon) caused only subadditive or additive effects on the LD_{50} values (Du Bois, 1961; West et al., 1961; Du Bois & Kinoshita, 1963; Frawley et al., 1963; Sanderson & Edson, 1964; McCollister et al., 1968; Flucke & Kimmerle (1977). Williams et al. (1957) found additive effects in their testing of oral combinations of methyl parathion with demeton, EPN, malathion, or ethyl parathion in dogs.

Mice pretreated with 50-300 mg diethyl dithiocarbamate per kg body weight displayed a remarkable reduction in the acute toxicity of methyl parathion. The toxicity was up to 10 times less. Lange & Wiezorek (1975) explained this observation by an effect of the dithiocarbamate on the microsomal oxidases and, thus, on the metabolism of methyl parathion. Another explanation is that compounds that temporarily occupy the active site of acetylcholinesterase prevent phosphorylation of the enzyme until there has been time for destruction of the organic phosphorus compound by A-esterases (Hayes & Laws, 1991).

Dithiocarb reduced the toxicity of methyl parathion in mice markedly, when applied 30 min before the methyl parathion. No effect was observed when dithiocarb and methyl parathion were applied simultaneously (Lange et al., 1977).

Orlando et al. (1972) found that pretreatment with quinidine sulfate had an inhibitory effect on the toxic action of i.v. injection of methyl parathion in rabbits. This could be demonstrated by electrocardiography. Quinidine sulfate reduced the influence of methyl parathion on the nicotinic-and muscarinic-type receptors.

The effect of methyl parathion on monoamine oxidase activity (MAO) in rat brain mitochondria was investigated by Nag & Nandi (1987). *In vitro* methyl parathion reduced the MAO significantly; however, *in vivo*, the effect was negligible.

Methyl parathion is an inhibitor of malate dehydrogenase in the mitochondria of liver and skeletal muscle. There was also an inhibitory effect on plasmatic malate dehydrogenase and lactate dehydrogenase in the liver (Tripathi & Shukla, 1988).

8.10 Mode of action

The mode of action of organophosphorus insecticides, such as methyl parathion, is described in Environmental Health Criteria 63 (WHO, 1986).

8.10.1 Inhibition of esterases

The primary biochemical effect associated with toxicity caused by organophosphorus pesticides is inhibition of acetylcholinesterase (AChE). The normal function of AChE is to terminate neuro-transmission due to acetylcholine, liberated at cholinergic nerve endings in response to nervous stimuli. Loss of AChE activity may lead to a range of effects resulting from excessive nervous stimulation and culminating in respiratory failure and death. The chemistry of the inhibition of AChE and of many other esterases (e.g., NTE and liver carboxyesterases, which are discussed elsewhere) by these chemicals is similar and is given in schematic form in Fig. 4. Following the formation of a Michaelis complex (reaction 1), a specific serine residue in the protein is phosphorylated with loss of the leaving group X (reaction 2). Two further reactions are possible: reaction 3 (reactivation) may occur spontaneously at a rate that is dependent on the nature of the attached group and on the protein and is also dependent on the influence of pH and of added nucleophilic reagents, such as oximes, which may catalyse reactivation. Reaction 4 ("aging") involves cleavage of an R-O-P-bond with the loss of R and the formation of a charged monosubstituted phosphoric acid residue still attached to protein. The reaction is called "aging" because it is time dependent, and the product is no longer responsive to nucleophilic reactivating agents, such as some oximes. Since therapy of organophosphorus compound poisoning is, in part, dependent on the reactivating power of oximes, understanding of the "aging" reaction is important. Pseudocholinesterase (ChE), which is present in blood plasma and nervous tissue, but has no known physiological function, is inhibited by organophosphorus compounds in a similar way to AChE, but the specificity of the 2 enzymes is different. Though no toxic effect arises as a result of inhibition of pseudoChE, measures of its inhibition can be made for monitoring purposes.

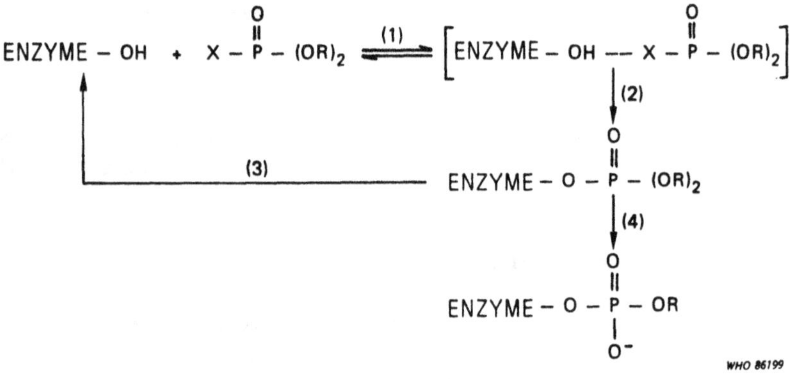

Fig. 4.   Inhibition of an esterase enzyme by organophosphorus compounds. From: WHO (1986).
(1)   Formation of Michaelis complex.
(2)   Phosphorylation of the enmzyme.
(3)   Reactivation reaction.
(4)   "Aging".

8.10.2 Possible alkylation of biological macromolecules

It has been shown, under laboratory conditions, that some organophosphates can react with, and alkylate, the reagent 4-nitro-benzylpyridine (Preussmann et al., 1969). The study was interpreted to imply that the *in vivo* alkylating potential of some pesticides was similar to that of the known mutagens, dimethyl sulfate and methyl methanesulfonate. Furthermore, Löfroth et al. (1969) derived a substrate constant (a logarithmic measure of alkylating ability) of

0.75 for dichlorvos, which is intermediate between those known for methyl and ethyl methanesulfonates. Concern over the possible mutagenic and carcinogenic potential of organophosphorus compounds on the basis of the above data was misplaced, since alternative reactions were not considered. Compared with the carbon atom of the alkyl group, the phosphorus atom is markedly more electron-deficient and susceptible to attack by nucleophiles. Analysis by Bedford & Robinson (1972) of the data of Löfroth et al. (1969) revealed that the proposed rates of alkylation by hard nucleophiles were probably combined rates of phosphorylation and alkylation, and that phosphorylation was the totally dominant reaction in the case of the hydroxide ion. The comparison with known mutagens was therefore inappropriate. Two factors detract further from the toxicological significance of the alkylation studies. The first is that mammalian tissues (plasma, liver, etc.) contain active enzymes that catalyse the phosphorylation of water by the organophosphorus esters. Viewed inversely, these enzymes (often called A-esterases) catalyse the hydrolysis of the organophosphorus esters, thereby rapidly reducing circulating levels of hazardous material. Secondly, the comparative rate of reaction of most of these pesticides with AChE is many orders greater than their rate of alkylation of the typical nucleophile 4-nitrobenzylpyridine: for dichlorvos, the ratio of rates was 1×10^7 in favour of the inhibitory phosphorylation of AChE (Aldridge & Johnson, 1977). It follows that, at low exposure levels, *in vivo* phosphorylation of AChE and other esterases will be the dominant reaction with negligible uncatalysed alkylation of genetic material. Indeed, no such alkylation has been detected in sensitive *in vivo* studies designed to check this point (Wooder et al., 1977). Some catalysed alkylations of glutathione by organosphorus compounds are known to occur *in vivo*, but these are essentially detoxification reactions.

8.10.3 General

Following lethal amounts of methyl parathion, hypotension, bradycardia, bronchoconstriction, and bronchial fluid accumulation occur with the inability of respiratory muscles to work. Cyanosis and central respiratory depression can be observed. In less severe cases of intoxication, bradycardia, muscle rigidity, muscle hypotonia, bronchial spasm, and constriction dominate (Meyer-Jones et al., 1977).

9. EFFECTS ON MAN

The only confirmed effects on humans of exposure to methyl parathion are the signs and symptoms characteristic of systemic poisoning by cholinesterase-inhibiting organophosphorus compounds, observed in case studies. The results of oral ingestion studies performed by Rider et al. (1969, 1970, 1971) suggest that manifestations of acute methyl parathion toxicity are absent in humans whose erythrocyte cholinesterase activity has been reduced to as little as 45% of their pre-exposure baselines (see section 9.1.2).

The effects of methyl parathion exposure on human beings were compiled in 1976 by NIOSH. Details are given by Hayes & Laws (1991).

WHO (1986) summarized the signs and symptoms of organo-phosphate insecticide poisoning as follows:

(a) Muscarinic manifestations

- increased bronchial secretion, excessive sweating, salivation, and lacrimation;
- pinpoint pupils, bronchoconstriction, abdominal cramps (vomiting and diarrhoea); and
- bradycardia.

(b) Nicotinic manifestations

- fasciculation of fine muscles and, in more severe cases, of the diaphragm and respiratory muscles; and
- tachycardia.

(c) Central nervous system manifestations

- headache, dizziness, restlessness, and anxiety;
- mental confusion, convulsions, and coma; and
- depression of the respiratory centre.

All these signs and symptoms can occur in different combinations and can vary in time of onset, sequence, and duration, depending on the chemical, dose, and route of exposure. Mild poisoning might include muscarinic and nicotinic signs only. Severe cases always show central nervous system involvement; the clinical picture is dominated by respiratory failure, sometimes leading to pulmonary oedema, due to the combination of the above-mentioned signs and symptoms.

9.1 General population exposure

The general population may be exposed to air-, water-, and food-borne residues of methyl parathion as a consequence of agricultural/forestry practices, the misuse of the agent, and contamination of field crops, water, and air by off-target loss.

Lisi et al. (1986, 1987) studied the allergic potential of methyl parathion in 200 persons. No significant sensitization to methyl parathion was found.

9.1.1 Acute toxicity

Several cases of methyl parathion poisoning have been reported throughout the world; these have been reviewed by Hayes & Laws (1991).

Human manifestations of acute poisoning by methyl parathion are comparable with those described in experimental animals (Durham & Hayes, 1962; Fazekas & Rengei, 1965; Hayes & Laws, 1991).

In cases of fatal methyl parathion poisoning, gross and microscopic alterations occur in all the organs (brain, lung, heart, liver, kidneys, spleen, vascular walls, perivascular areas). Fazekas (1971) already saw alterations due to methyl parathion-poisoning, 2 h after the poisoning. Ember et al. (1970) found a high content of vitamin A in the liver in 5 cases of suicide with methyl parathion.

Van Bao et al. (1974) reported an increase in chromosome aberrations in the lymphocytes of 4 patients who had suffered acute methyl parathion poisoning as a result of attempted suicide. The increase in chromosome aberrations was detected only in cell cultures carried out 1 month after their admission to hospital. No significant changes were found, compared with controls, 6 months later.

9.1.2 *Effects of short- and long-term exposure, controlled human studies*

Five male volunteers received 3.0 mg methyl parathion per day for 28 days, then 3.5 mg methyl parathion for 28 days, and 4.0 mg methyl parathion for 43 days. No symptoms of poisoning or effects on the plasma or red blood cell cholinesterases could be noticed (Moeller & Rider, 1961).

In another study, 3 groups of 5 volunteers each received 4.5 mg methyl parathion daily, for 30 days, then 5.0 g for 29 days or 5.5 mg for 28 days, followed by 6.0 mg for 29 days or 6.5 mg for 35 days, and finally 7.0 mg for 24 days. In no case was significant inhibition of the plasma or red blood cell cholinesterase activity found (Moeller & Rider, 1962).

Morgan et al. (1977) studied the cholinesterase activities in 4 human volunteers, who received 2 or 4 mg methyl parathion on 5 successive days. These doses did not cause any depression of plasma and red blood cell cholinesterase activity.

Rider et al. (1969) reported studies on human volunteers to determine the level of minimal toxicity of methyl parathion. For 30 days, 5 volunteers received capsules containing methyl parathion, with the dose increasing daily, and 2 received capsules containing corn oil. Depression in plasma cholinesterase activity (15%) was observed at an oral dosage of 11.0 mg per day, while, at higher dosages up to and including 19 mg per day, no significant cholinesterase depression was observed. No significant changes in the blood cell count, urine analysis, or the prothrombin times occurred, nor was there any evidence of toxic effects.

After 4 weeks with daily doses of 24 mg methyl parathion, 2 out of 5 volunteers showed inhibition of plasma and red blood cell cholinesterase activities with decreases of 24 or 23% for plasma and 27 or 55% for red blood cells (Rider et al., 1970).

Five volunteers received doses increasing from 14 to 20 mg methyl parathion per day, orally, for 6 days. No inhibition of the cholinesterases was found. However, doses of 28 or 30 mg methyl parathion caused a decrease in the cholinesterase activities of about 37% (Rider et al., 1971).

Two male volunteers received, orally, 2 or 4 mg methyl parathion per day. No influence of methyl parathion on neurophysiological parameters was found, and there was no inhibition of the plasma or red blood cell cholinesterase activity (Rodnitzky et al., 1978).

9.2 Occupational exposure

The production, formulation, handling, and use as an insecticide of methyl parathion are potential sources of exposure. Skin contact or inhalation are the main hazards for workers. The main hazard for the general population is the ingestion of contaminated food. Winddrift during spraying may be a health risk, since Kummer & Van Sittert (1986) observed that, in a number of cases, the spraymen did not stop spraying, when it was too windy.

The analysis of 375 pesticide poisonings in Bulgaria during 1965-68 showed that 82.5% of all cases were due to organophosphates. Six of the intoxications were attributed to methyl parathion. A large number of poisonings, usually mild, occurred not in applicators directly engaged in plant protection but in other agricultural workers when they entered a previously sprayed crop area for further cultivation and hand-harvesting (Kaloyanova-Simeonova, 1970).

Hatcher & Wiseman (1969) reported 16 cases of methyl parathion intoxication among 118 organophosphorus insecticide poisonings of farm workers that occurred in the lower Rio Grande Valley (Texas) in 1968. Toxicity following dermal exposure was predominant.

Neuropsychiatric sequelae from occupational exposure to organophosphorus pesticides have been reported (Dille & Smith, 1964). However, the patients had been exposed to other pesticides besides methyl parathion.

Data on chromosomal aberrations due to methyl parathion are scarce. Data from persons who had worked with various pesticides were presented by Yoder et al., 1973 (positive finding); Rupa et al., 1989 (positive finding); and Nehéz et al., 1988 (positive finding in farm workers in the open field, but not in those in enclosed spaces like greenhouses). Van Bao et al. (1974) found chromosome aberrations in one case of an agricultural worker, accidentally exposed to methyl parathion (without exposure data).

De Cassia Stocco et al. (1982) reported data from subjects exposed to methyl parathion and DDT at a formulation plant near of Sao Paulo, Brazil. No increased frequency of chromosome aberrations was found in the lymphocyte cultures of 15 healthy male workers (with blood cholinesterase level ≤ 75% of presumably the normal mean levels), who were exposed repeatedly or long-term to methyl parathion for durations ranging from 1 week to 7 years, but who had intermittent periods of non-exposure.

Richter et al. (1986) investigated the risk of exposure to methyl parathion spray drift in the workers in 3 kibbutzim. The cholinesterase levels were measured in 36 agricultural workers and 25 residents from the same kibbutzim. No effects due to the methyl parathion spray drift exposure were observed in the field workers or in the residents.

9.2.1 Epidemiological studies

There are no epidemiological studies on effects related only to methyl parathion exposure.

10. PREVIOUS EVALUATIONS BY INTERNATIONAL BODIES

The FAO/WHO Joint Meeting on Pesticide Residues (JMPR) evaluated methyl parathion in 1968, 1972, 1975, 1979, 1980, and 1984 (FAO/WHO, 1969, 1973, 1976, 1980, 1981 and 1985). The acceptable daily intake for man (ADI) was estimated at 0-0.02 mg/kg body weight in 1984. This was based on levels causing no toxicological effects of:

- 2 mg/kg diet, equivalent to 0.1 mg/kg body weight in the rat; and

- 0.3 mg/kg body weight per day in man.

The FAO/WHO Codex Alimentarius Commission (FAO/WHO, 1986) recommended Maximum Residue Limits (MRLs) in several food commodities, ranging from 0.05 to 0.2 mg/kg as follows:

Commodity	MRL (mg/kg)
Cantaloupe	0.2
Cole crops	0.2
Cottonseed oil	0.05
Cucumbers	0.2
Fruit, other	0.2
Hops (dry cones)	0.05[a]
Melons	0.2
Sugar beets	0.05[a]
Tea (fermented and dried)	0.2
Tomatoes	0.2

[a] Levels at, or about, the limit of determination.

The International Agency for Research on Cancer (IARC) evaluated methyl parathion in 1982 and in 1987 (IARC, 1983, 1987), and concluded that the available data do not provide evidence that methyl parathion is carcinogenic to experimental animals. No data on humans were available. The available data provide no evidence that methyl parathion is likely to present a carcinogenic risk for humans.

WHO (1990) classified technical methyl parathion as "extremely hazardous" in normal use, based on an oral LD_{50} in the rat of 14 mg/kg. WHO/FAO (1975) issued a data sheet on methyl parathion (No. 7).

REFERENCES

ABE, T., FUJIMOTO, Y., TATSUNO, T., & FUKAMI, J. (1979) Separation of methyl parathion and fenitrothion metabolites by liquid chromatography. Bull. environ. Contam. Toxicol., 22: 791-795.

ACKERMANN, H. (1974) [Transfer of organophosphorous insecticides to the embryo-formation of PO derivatives and development of toxic symptoms.] Tagungsber. Akad. Landwirtsch., 126: 23-29 (in German).

ADHYA, T.K., SUDHAKAR-BARIK, & SETHUNATHAN, N. (1981a) Stability of commercial formulation of fenitrothion, methyl parathion, and parathion in anaerobic soils. J. Agric. Food Chem., 29: 90-93.

ADHYA, T.K., SUDHAKAR-BARIK, & SETHUNATHAN, N. (1981b) Fate of fenitrothion, methyl parathion, and parathion in anoxic sulfur-containing soil systems. Pest. Biochem. Physiol., 16: 14-20.

ADHYA, T.K., SUDHAKAR-BARIK, & SETHUNATHAN, N. (1981c) Hydrolysis of selected organophosphorus insecticides by two bacteria isolated from flooded soil. J. appl. Bacteriol., 50: 167-172.

ADHYA, T.K., WAHID, P.A., & SETHUNATHAN, N. (1987) Persistence and biodegradation of selected organophosphorus insecticides in flooded versus non-flooded soils. Biol. Fertil. Soils, 4(1): 36-40.

AGISHEV, M.K., TSUKERMAN, V.G., & KUTSENKO, A.A. (1977) [Methods for determining pesticide residues in soil.] Khim. sel'sk. Khoz., 15: 52-54 (in Russian).

AGOSTIANO, A., CASELLI, M., & PROVENZANO, M.R. (1983) Analysis of pesticides and other organic pollutants by preconcentration and chromatographic techniques. Water Air Soil Pollut., 19: 309-320.

AKHMEDOV, B.K. (1968) The hygienic significance of metaphos as an atmospheric pollutant. Gig. i Sanit., 33: 10-15.

ALBANIS, T.A., POMONIS, P.J., & SDOUKOS, A.T. (1986) Organophosphorus and carbamate pesticide residues in the aquatic system of Ioannina basin and Kalamas River (Greece). Chemosphere, 15: 1023-1034.

ALBAUGH, D.W. (1972) Insecticide tolerances of two crayfish populations (Procambarus acutus) in south-central Texas. Bull. environ. Contam. Toxicol., 8(6): 334-338.

ALBERT, L., GONZALEZ, M., & MARTINEZ-DEWANE, M.G. (1979) [Organophosphorus insecticides. I. Residues of organophosphorus insecticides in some Mexican foods.] Rev. Soc. Quim. Méx., 23(4): 189-197 (in Spanish).

ALDRIDGE, W.N. & JOHNSON, M.K. (1977) Mechanisms and structure activity relationships providing a high safety factor for anti-cholinesterase insecticides. In: Proceedings of the British Crop Protection Conference, Brighton, November 1977, Croydon, British Crop Protection Council, pp. 721-729.

AMBRUS, A., HARGITAL, E., KAROLY, G., FULOP, A., & LANTOS, J. (1981a) General method for determination of pesticide residues in samples of plant origin, soil, and water. II. Thin layer chromatographic determination. J. Assoc. Off. Anal. Chem., 64: 743-748.

AMBRUS, A., VISI, E., ZAKAR, F., HARGITAI, E., SZABO, L., & PAPA, A. (1981b) General method for determination of pesticide residues in samples of plant origin, soil, and water. III. Gas chromatographic analysis and confirmation. J. Assoc. Off. Anal. Chem., **64**: 749-768.

ANDERSON, J.F. & WOJTAS, M.A. (1986) Honey bees (Hymenoptera: Apidae) contaminated with pesticides and polychlorinated byphenyls. J. Econ. Entomol., **79**: 1200-1205.

ANDERSSON, A. (1986) Monitoring and biased sampling of pesticide residues in fruits and vegetables. Methods and results, 1981-1984. Var Föda, **38**(Suppl. 1): 8-55.

ANDERSSON, A. & OHLIN, B. (1986) A capillary gas chromatographic multiresidue method for determination of pesticides in fruits and vegetables. Var Föda, **38**(Suppl. 2): 79-109.

ANEES, M.A. (1975) Acute toxicity of four organophosphorus insecticides to a freshwater teleost *Channa punctatus* (Bloch). Pak. J. Zool., **7**: 135-141.

ANON (1984) Hazardous chemical data: Vol. 2. Chris. Washington, DC, US Department of Transportation.

APPERSON, C.S., ELSTON, R., & CASTLE, W. (1976) Biological effects and persistence of methyl parathion in Clear Lake, California. Environ. Entomol., **5**: 1116-1120.

APPERSON, C.S., YOWS, D., & MADISON, C. (1978) Resistance to methyl parathion in *Chaoborus astcoptus*. J. Econ. Entomol., **71**(5): 772-773.

ARCHER, T.E. & ZWEIG, G. (1959) Direct colorimetric analysis of cholinesterase-inhibitions with indophenyl acetate. J. Agric. Food Chem., **7**: 1-78.

ARCHER, S.R., McCURLEY, W.R., & RAWLINGS, G.D. (1978) Source assessment: pesticide manufacturing air emissions - overview and prioritization. Washington, DC, US Environmental Protection Agency, p. 135 (EPA-600/2-78-004d).

ARNDT, W., SCHINDLBECK, E., PARLAR, H., & KORTE, F. (1981) [Reaction of the organophosphate insecticides during rubbish composting.] Chemosphere, **10**: 1035-1040 (in German).

ARTERBERRY, J.D., DURHAM, W.F., ELLIOT, J.W., & WOLFE, H.R. (1961) Exposure to parathion. Arch. environ. Health, **3**: 112-121.

ARTHUR, R.D., CAIN, J.D., & BARRANTINE, B.F. (1976) Atmospheric levels of pesticides in Mississippi delta. Bull. environ. Contam. Toxicol., **15**: 129-134.

AUDEGOND, L., CATEZ, D., FOULHOUX, P., FOURNEX, R., LE RUMEUR, C., L'HOTELLIER, M., & STEPNIEWSKI, J.P. (1989) Potentialisation de la toxicité de la deltaméthrine par les insecticides organophosphorés. J. Toxicol. clin. Exp., **9**: 163-176.

AULERICH, R.J., RINGER, R.K., & SAFRONOFF, J. (1987) Primary and secondary toxicity of warfarin, sodium monofluoroacetate, and methyl parathion in mink. Arch. environ. Contam. Toxicol., **16**: 357-366.

AULT, J.A., SCHOFIELD, C.M., JOHNSON, L.D., & WALTZ, R.H. (1979) Automated gel permeation chromatographic preparation of vegetables, fruits, and crops for organophosphate residue determination utilizing flame photometric detection. J. Agric. Food Chem., 27: 825-828.

BABU, S.B.R., JAYASUNDARAMMA, B., & RAMAMURTHI, R. (1986) Behavioral abnormalities of juveniles of the fish *Cyprinus carpio* exposed to methyl parathion. Environ. Ecol., 4: 95-97.

BADAWY, M.I. & EL-DIB, M.A. (1984) Persistence and fate of methyl parathion in sea water. Bull. environ. Contam. Toxicol., 33: 40-49.

BADAWY, M.I., EL-DIB, M.A. & OSAMA, A.A. (1984) Spill of methyl parathion in the Mediterranean Sea: A case study at Port-Said, Egypt. Bull. environ. Contam. Toxicol., 32: 469-477.

BAKER, R.D. & APPLEGATE, H.G. (1970) Effect of temperature and ultraviolet radiation on the persistence of methyl parathion and DDT in soils. Agron. J., 62: 509-512.

BAKER, R.D. & APPLEGATE, H.G. (1974) Effect of ultraviolet radiation on the persistence of pesticides. Texas J. Sci., 5: 53-59.

BARNES, J.M. & DENZ, F.A. (1953) Experimental demyelination with organo-phosphorous compounds. J. Pathol. Bacteriol., 65: 597-605.

BECKER, G. (1971) [Simultaneous gas chromatographic determination of chlorinated hydrocarbons and phosphates in plant material.] Dtsch. Lebensm.- Rundsch., 67: 125-126 (in German).

BECKER, G. (1979) [Multimethod for simultaneous detection of 75 plant treatment preparations on plant material.] Dtsch. Lebensm.-Rundsch., 75: 148-152 (in German).

BECKER, G. & SCHUG, P. (1990) [A micromethod for the rapid determination of chlorinated hydrocarbons and phosphate esters in plant-derived foods.] Dtsch. Lebensm. Rundsch., 86(8): 239-242 (in German).

BECKMAN, H. & GARBER, D. (1969) Pesticide residues. Recovery of 65 organophosphorus pesticides from Florisil with a new solvent-elution system. J. Assoc. Off. Anal. Chem., 52: 286-293.

BEDFORD, C.T. & ROBINSON, J. (1972) The alkylating properties of organophosphates. Xenobiotica, 2: 307-337.

BEINE, H. (1987) [Phosphoric acid esters and related compounds - environmental significance and determination in air.] Essen, Immission Protection Institute for the Land of North Rhine-Westphalia, 31 pp. (LIS Report No., 69) (in German with English summary).

BELASHOVA, I.G., KHOKHOL'KOVA, G.A., & KLISENKO, M.A. (1983) [Selection of environmental air sampler containing pesticide vapors.] Gig. Tr. Prof. Zabol., 54-56 (in Russian).

BELISLE, A.A. & SWINEFORD, D.M. (1988) Simple specific analysis of organophosphorus and carbamate pesticides in sediments using column extraction and gas chromatography. Environ. Toxicol. Chem., 7: 749-752.

BENES, V. & JELINEK, R. (1979) Embryotoxicity of parathion methyl with the Chick Embryotoxicity Screening Test (CHEST). In: Pesticide residues in food: 1979 evaluations - The monographs. Rome, Food and Agriculture Organization of the United Nations, pp. 367-368 (FAO Plant Production and Protection Paper 20, suppl.).

BENKE, G.M. & MURPHY, S.D. (1975) The influence of age on the toxicity and metabolism of methyl parathion and parathion in male and female rats. Toxicol. appl. Pharmacol., 31: 254-269.

BENKE, G.M., CHEEVER, K.L., MIRER, F.E., & MURPHY, S.D. (1974) Comparative toxicity, anticholinesterase action and metabolism of methyl-parathion and parathion in sunfish and mice. Toxicol. appl. Pharmacol., 28(1): 97-109.

BENNETT, R.S. (1989) Role of dietary choices in the ability of bobwhite to discriminate between insecticide-treated and untreated food. Environ. Toxicol. Chem., 8: 731-738.

BENNETT, R.S., WILLIAMS, B.A., SCHMEDDING, D.W., & BENNETT, J.K. (1991) Effects of dietary exposure to methyl parathion on egg laying and incubation in mallards. Environ. Toxicol. Chem., 10: 501-507.

BETOWSKI, L.D. & JONES, T.L. (1988) Analysis of organophosphorus pesticide samples by high-performance liquid chromatography/mass spectrometry and high-performance liquid chromatography/mass spectrometry/mass spectrometry. Environ. Sci. Technol., 22: 1430-1434.

BHASKAR, S.U. & KUMAR, N.V.N. (1981) Thin layer chromatographic determination of methyl parathion as paraoxon by cholinesterase inhibition. J. Assoc. Off. Anal. Chem., 64: 1312-1314.

BHUNIA, A.K, BASU, N.K., ROY, D., CHAKRABARTI, A., & BANERJEE, S.K. (1991) Growth, chlorophyll content, nitrogen-fixing ability, and certain metabolic activities of *Nostoc muscorum*: effect of methyl parathion and benthiocarb. Bull. environ. Contam. Toxicol., 47: 43-50.

BHASKAR, S.U. & KUMAR, N.V.N. (1982) Selection of enzyme sources for improved sensitivity in enzymic determination of organophosphorus pesticides. J. Assoc. Off. Anal. Chem., 65: 1297-1298.

BHASKAR, S.U. & KUMAR, N.V.N. (1984) A rapid colorimetric method for determination of methyl parathion residues. Pesticides, 18: 17-18.

BIGLEY, W.S., PLAPP, F.W. Jr, HANNA, R.L., & HARDING, J.A. (1981) Effect of toxaphene, camphene, and cedar oil on methyl parathion residues on cotton. Bull. environ. Contam. Toxicol., 27: 90-94.

BLANCO, C.C. & SANCHEZ, F.G. (1989) Derivative spectrophotometric determination of the organophosphate insecticide methyl parathion in commercial formulations. Analusis, 17(1-2): 80-86.

BOMHARD, E., LOSER, E., & SCHILDE, B. (1981) [E 605-methyl (methyl parathion). Chronic toxicological studies in rats (2-year feeding trial).] Wuppertal-Elberfeld, Bayer

AG, Institute of Toxicology (Unpublished Report No. 9889, submitted to WHO by Bayer AG, Leverkusen, Germany) (in German).

BOURQUE, C.L., DUGUAY, M.M., & GAUTREAU, Z.M. (1989) The determination of reducible pesticides by adsorptive stripping voltammetry. Int. J. environ. Stud., **37**: 187-197.

BOURQUIN, A.W., GARNAS, R.L., PRITCHARD, P.H., WILKES, F.G., CRIPE, C.R., & RUBINSTEIN, N.I. (1979) Interdependent microsomes for the assessment of pollutants in the marine environment. Int. J. environ. Stud., **13**: 131-140.

BOWMAN, M.C. (1981) Analysis of organophosphorus pesticides. In: Moye H.A., ed. Analysis of pesticide residues, New York, Basel, Marcel Dekker, pp. 263-332.

BOWMAN, M.C. & BEROZA, M. (1967) Temperature-programmed gas chromatography of 20 phosphorus-containing insecticides on 4 different columns and its application to the analysis of milk and corn silage. J. Assoc. Off. Anal. Chem., **50**: 1228-1236.

BOWMAN, B.T. & SANS, W.W. (1979) The aqueous solubility of twenty-seven insecticides and related compounds. J. Environ. Sci. Health, **B14**: 625-634.

BRAECKMAN, R.A., GODEFROOT, M.G., BLONDEEL, G.M., & BELPAIRE, F.M. (1980) Kinetic analysis of the fate of methyl parathion in the dog. Arch. Toxicol., **43**: 263-271.

BRAECKMAN, R.A., AUDENAERT, F., WILLEMS, J.L., BELPAIRE, F.M., & BOGAERT, M.G. (1983) Toxicokinetics of methyl parathion and parathion in the dog after intravenous and oral administration. Arch. Toxicol., **54**: 71-82.

BRANCA, P. & QUAGLINO, P. (1988) [Presence of antiparasite pesticides in imported potatoes.] Ind. Aliment., **27**: 427-431 (in Italian).

BRATIN, K., KISSINGER, P.T., & BRUNTLETT, C.S. (1981) Reductive mode thin-layer amperometric detector for liquid chromatography. J. Liq. Chromatogr., **4**: 1777-1795.

BREWER, L.W., DRIVER, C.J., KENDALL, R.J., ZENIER, C., & LACHER, T.E. Jr. (1988) Effects of methyl parathion in ducks and duck broods. Environ. Toxicol. Chem., **7**: 375-379.

BRODESSER, J. & SCHOELER, H.F. (1987) An improved extraction method for the quantitative analysis of pesticides in water. Zbl. Bakteriol. Mikrobiol. Hyg. Ser. B, **185**: 183-185.

BRODEUR, J. & DU BOIS, K.P. (1963) Comparison of acute toxicity of anticholinesterase insecticides to weanling and adult male rats. Proc. Soc. Exp. Biol. Med., **114**: 509-511.

BROWN, L.C., CATHEY, G.W., & LINCOLN, C. (1962) Growth and development of cotton as affected by toxaphene-DDT, methyl parathion, and calcium arsenate. J. Econ. Entomol., **55**: 298-301.

BRUCHET, A., COGNET, L., & MALLEVIALLE, J. (1984) Continuous composite sampling and analysis of pesticides in water. Water Res., **18**: 1401-1409.

BUBIEN, Z. (1971) [Intoxication of animals with phosphoorganic insecticides.] Med. Wet., 27(10): 611-612 (in Polish with English summary).

BUCK, N.A., ESTESEN, B.J., & WARE, G.W. (1980) Dislodgeable insecticide residues on cotton foliage: fenvalarate, permethrin, sulprofos, chlorpyrifos, methyl parathion, EPN, oxamyl, and profenofos. Bull. environ. Contam. Toxicol., 24: 283-288.

BUERGER, T.T, KENDALL, R.J., MUELLER, B.S., DE VOS, T., & WILLIAMS, B.A. (1991) Effects of methyl parathion on northern bobwhite survivability. Environ. Toxicol. Chem., 10: 527-532.

BUTLER, P.A. (1964)   Commercial fishery investigation. Washington, DC, US Department of the Interior, Fish and Wildlife Service, pp. 5-28 (Circular No. 199).

BUTLER, P.A. & SCHUTZMANN, R.L. (1978)   Residues of pesticides and PCBs in estuarine fish, 1972-1976. National Pesticide Monitoring Program. Pestic. Monit. J., 12: 51-59.

BUTLER, L.C., STAIFF, D.C., & DAVIES, J.E. (1981) Methyl parathion persistence in soil following simulated spillage. Arch. environ. Contam. Toxicol., 10: 451-458.

CAPRIEL, P, HAISCH, A., & SHAHAMAT, U. (1986) Supercritical methanol: an efficacious technique for the extraction of bound pesticide residues from soil and plant samples. J. Agric. Food Chem., 34: 70-73.

CARERE, A., ORTALI, V.A., CARDAMONE, G., & MORPURGO, G. (1978) Mutagenicity of dichlorvos and other structurally related pesticides in *Salmonella* and *Streptomyces*. Chem.-Biol. Interact., 22: 297-308.

CAREY, A.E. & KUTZ, F.W.   (1985)   Trends in ambient concentrations of agrochemicals in humans and the environment of the United States.   Environ. Monit. Assess., 5: 155-163.

CAREY, A.E., GOWEN, J.A., TAI, H., MITCHELL, W.G., & WIERSMA, B. (1978) Soils, pesticide residue levels in soils and crops, 1971 - National Soils Monitoring Program (III). Pestic. Monit. J., 12: 117-136.

CAREY, A.E., GOWEN, J.A., TAI, H., MITCHELL, W.G., & WIERSMA, G.B. (1979) Pesticide residue levels in soils and crops from 37 states, 1972. National Soils Monitoring Program (IV). Pestic. Monit. J., 12: 209-229.

CHAMBERS, J.E. & YARBROUGH, J.D. (1974)   Parathion and methyl-parathion toxicity to insecticide-resistant and susceptible mosquitofish *Gambusia affinis*. Bull. environ. Contam. Toxicol., 11(4): 315-320.

CHANG, C.S. & LANGE, W.H. (1967) Laboratory and field evaluation of selected pesticides for control of the red crayfish in California rice fields. J. Econ. Entomol., 60(21): 473-477.

CHAUDHRY, G.R., ALI A.N., & WHEELER, W.B. (1988) Isolation of a methyl parathion-degrading *Pseudomonas* sp. that possesses DNA homologous to the opd gene from a *Flavobacterium* sp. Appl. environ. Microbiol., 54(2): 288-293.

CHEAH, M.L., AVAULT, J.W. Jr, & GRAVES, J.B. (1980) Some effects of rice pesticides on crawfish. Louisiana. Agric., 23(2): 8-9, 11.

CHEMICAL INFORMATION SERVICES, LTD (1988) Directory of world chemical producers: 1989/90 edition. Oceanside, New York, Chemical Information Services, Ltd, Publisher.

CHEN, H.H., HSUEH, J.L., SIRIANNI, S.R., & HUANG, C.C. (1981) Induction of sister-chromatid exchanges and cell cycle delay in cultured mammalian cells treated with eight organophosphorus pesticides. Mutat. Res., 88: 307-316.

CHEN, Z., ZABIK, M.J., & LEAVITT, R.A. (1984) Comparative study of thin film photodegradative rates for 36 pesticides. Ind. Eng. Chem. Prod. Res. Dev., 23: 5-11.

CHERNYAK, S.M. & ORADOVSKII, S.G. (1980) [Determination of organochlorine and organophosphorus pesticides in sea water by gas-liquid chromatography.] Metody Issled. Org. Veshchestva Okeane, 1980: 291-303 (in Russian).

CHMIL, V.D., GRAHL, K., & STOTTMEISTER, E. (1978) Chromatographic methods of determining residual amounts of pesticides in water. Zh. Anal. Khim., 33: 2420-2425.

CHUKWUDEBE, A., MARCH, R.B., OTHMAN, M., & FUKUTO, T.R. (1989) Formation of trialkyl phosphorothioate esters from organophosphorus insecticides after exposure to either ultraviolet light or sunlight. J. Agric. Food Chem., 37: 539-545.

CLARK, D.R. Jr (1986) Toxicity of methyl parathion to bats: mortality and coordination loss. Environ. Toxicol. Chem., 5(2): 191-195.

CLARK, G.J., GOODIN, R.R., & SMILEY, J.W. (1985) Comparison of ultraviolet and reductive amperometric detection for the determination of ethyl and methyl parathion in green vegetables and surface water using high-performance liquid chromatography. Anal. Chem., 57: 2223-2228.

COBURN, J.A. & CHAU, A.S.Y. (1974) Confirmation of pesticide residue identity. VIII. Organophosphorus pesticides. J. Assoc. Off. Anal. Chem., 57: 1272- 1278.

COHEN, M.L. & STEINMETZ, W.D. (1986) Foliar washoff of pesticides by rainfall. Environ. Sci. Technol., 20: 521-523.

COLE, C.L., MCCASLAND, W.E., & DACUS, S.C. (1986) The persistence of selected insecticides used in water and in water-oil sprays as related to worker re-entry. Southwest. Entomol., 11(Suppl.): 83-87.

COMPTON, B. (1973) Analysis of pesticides in air. Prog. Anal. Chem., 5: 133-152.

CORNELIUSSEN, P.E. (1970) Residues in food and feed. Pesticide residues in total diet samples (V). Pestic. Monit. J., 4: 89-105.

COSTA, L.G. & MURPHY, S.D. (1984) Interaction between acetaminophen and organophosphates in mice. Res. Commun. Chem. Pathol. Pharmacol., 44: 389-400.

COWART, R.P, BONNER, F.L., & EPPS, E.A. Jr (1971) Rate of hydrolysis of seven organophosphate pesticides. Bull. environ. Contam. Toxicol., 6: 231- 234.

CRIPE, G.M., NIMMO, D.R., & HAMAKER, T.L. (1981) Effects of two organophosphate pesticides on swimming stamina of the mysid *Mysidopsis bahia*. In: Vernberg, J., Calabrese, A., Thurberg, F.P., & Vernberg, W.B., ed. Biological monitoring of marine pollutants. New York, London, San Francisco, Academic Press, pp. 21-36.

CRIPE, C.R., WALTER, W.W., PRITCHARD, P.H., & BOURQUIN, A.W. (1987) A shake-flask test for estimation of biodegradability of toxic organic substances in the aquatic environment. Ecotoxicol. environ. Saf., 14(3): 239-251.

CROFT, B.A. (1977) Resistance in arthropod predators and parasites. In: Watson, D.L. & Brown, A.W.A., ed. Pesticide management and insecticide resistance. New York, London, San Francisco, Academic Press, pp. 377-393.

CROSSLAND, N.O. & BENNETT, D. (1984) Fate and biological effects of methyl parathion in outdoor ponds and laboratory aquaria, I: Fate. Ecotoxicol. Environ. Saf., 8: 471-481.

CROSSLAND, N.O. & ELGAR, K.E. (1983) Fate and biological effects of insecticides in ponds. In: Matsunaka, S., Hutson, D.H., & Murphy, S.D., ed. IUPAC. Pesticide chemistry: human welfare and the environment. Proceedings of the 5th International Congress of Pesticide Chemistry, Kyoto, Japan, 29 August-4 September 1982. Volume 3: Mode of action, metabolism and toxicology, Oxford, New York, Pergamon Press, pp. 551-556.

CROSSLAND, N.O., BENNETT, D., & WOLFF, C.J.M. (1986) Evaluation of models used to assess the fate of chemicals in aquatic systems. Pestic. Sci., 17: 297-304.

CSERNATONI, M., GYORFI, L., & HARGITAL, E. (1988) Results of pesticide analyses monitoring programme in surface waters in Hungary between 1977 and 1986. In: Abbou, R., ed. Hazardous waste: detection, control, treatment. Amsterdam, Oxford, New York, Elsevier Science Publishers, pp. 861-868.

CURINI, M., LAGANA, A., PETRONIO, B.M., & RUSSO, M.V. (1980) Determination of organophosphorus pesticides by thin-layer chromatography. Talanta, 27: 45-48.

CUSTER, T.W. & MITCHELL, C.A. (1987) Exposure to insecticides of brushland wildlife within the Lower Rio Grande Valley, Texas, USA. Environ. Pollut., 45(3): 207-220.

DALDRUP, T., SUSANTO, F., & MICHALKE, P. (1981) [Combination of TLC, GLC (OV 1 and OV 17) and HPLC (RP 18) for a rapid detection of drugs.] Fresenius Z. Anal. Chem., 308: 413-427 (in German).

DALDRUP, T., MICHALKE, P., & BOEHME, W. (1982) A screening test for pharmaceuticals, drugs and insecticides with reversed-phase liquid chromatography - retention data of 560 compounds. Chromatogr. Newsl., 10: 1-7.

DALY, I.W.A. (1983) Evaluation report for US-EPA of a two-year chronic feeding study of methyl parathion in rats. East Millstone, New Jersey, Biodynamics Inc. (Unpublished report, submitted to WHO by Bayer AG, Leverkusen, Germany).

DAMICO, J.N., BARRON, R.P, & SPHON, J.A. (1969) Field ionization spectrum of some pesticidal and other biologically significant compounds. Int. J. mass Spectrom. ion Phys., 2: 161-182.

DANKA, R.G., RINDERER, T.E., HELLMICH, II, R.L., & COLLINS, A.M. (1986) Comparative toxicities of four topically applied insecticides to Africanized and European Honey Bees (Hymenoptera: Apidae). J. Econ. Entomol., **79**: 18-21.

DAVIDSON, J.H., RAO, P.S.C, OU, L.T., WHEELER, W.B., & ROTHWELL, F.F. (1980) Adsorption, movement, and biological degradation of large concentrations of selected pesticides in soils. Gainesville, Florida, Florida University.

DAVIS, J.E., STAIFF, D.C., BUTLER, L.C., & STEVENS, E.R. (1981) Potential exposure to dislodgeable residues after application of two formulations of methyl parathion to apple trees. Bull. environ. Contam. Toxicol., **27**: 95-100.

DEAN, B.J. (1972) The mutagenic effects of organophosphorus pesticides on micro-organisms. Arch. Toxicol., **30**: 67-74.

DE CASSIA STOCCO, R., BECAK, W., GAETA, R., & NAZARETH RABELLOGAY, M. (1982) Cytogenetic study of workers exposed to methylparathion. Mutat. Res., **103**: 71-76.

DEGRAEVE, N. & MOUTSCHEN, J. (1984) Absence of genetic and cytogenetic effects in mice treated by the organophosphorus insecticide parathion, its methyl analogue, and paraoxon. Toxicology, **32**: 177-183.

DEGRAEVE, N., CHOLLET, M.C., & MOUTSCHEN, J. (1984) Cytogenic and genetic effects of subchronic treatment with organophosphorus insecticides. Arch. Toxicol., **56**: 66-73.

DEICHMANN, W.B., PUGLIESE, W., & CASSIDY, B.S. (1952) Effects of dimethyl and diethyl paranitrophenyl thiophosphate on experimental animals. Ind. Hyg. occup. Med., **5**: 44-51.

DELI, E. & KISS, Z. (1986) The effect of organophosphorous insecticide Wofatox 50 EC on the adenylate cyclase activity of chicken embryo muscle. Biochem. Pharmacol., **35**: 1603-1605.

DELI, E. & KISS, Z. (1988) Effect of parathion and methylparathion on protein content of chicken embryo muscle *in vivo*. Biochem. Pharmacol., **37**: 3251- 3256.

DELI, E. & VARNAGY, L. (1985) Teratological examination of Wofatox 50 EC (50% methyl parathion) on pheasant embryos. Anat. Anz. Jena, **158**: 237-240.

DELI, E., SOMLYAY, I., & VARNAGY, L. (1985) Biochemical study of muscle samples from chicken embryos affected by Wofatox 50 EC. Arch. Toxicol, **8**: 277-279.

DE POTTER, M., MULLER, R., & WILLEMS, J. (1978) A method for the determination of some organophosphorus insecticides in human serum. Chromatographia, **11**: 220-222.

DE SCHRYVER, E., DE REU, L., BELPAIRE, F., & WILLEMS, J. (1987) Toxicokinetics of methyl paraoxon in the dog. Arch. Toxicol., **59**: 319-322.

DESHPANDE, A.A. & SWAMY, G.S. (1987) Induction of proline accumulation by methyl parathion in sorghum (*Sorghum bicolor* L.). Curr. Sci., **56**: 1068-1070.

DEVI, Y.P., KUMAR, N.V., & HARAN, N.V.H. (1982) Survey of pesticide residue analysis in water resources and paddy straw samples from some farms of the Nellore district of Andhra Pradesh. Pollut. Res., 1: 21-23.

DE WIT, J.S.M., PARKER, C.E., TOMER, KB., & JORGENSON, J.W. (1987) Direct coupling of open-tubular liquid chromatography with mass spectrometry. Anal. Chem., 59: 2400-2404.

DICK, G.L., HEENAN, M.P., LOVE, J.L., UDY, P.B., & DAVIDSON, F. (1978) Survey of trace elements and pesticide residues in the New Zealand diet. 2. Organochlorine and organophosphorus pesticide residue content. N. Z. J. Sci., 21: 71-78.

DILLE, J.R. & SMITH, P.W. (1964) Central nervous system effects of chronic exposure to organophosphate insecticides. Aerosp. Med., 35: 475-478.

DI MUCCIO, A., AUSILI, A., VERGORI, L., CAMONI, I., DOMMARCO, R. & GAMBETTI, L. (1990) Single step multi-cartridge clean-up for organophosphate pesticide residue determination in vegetable oil extracts by gas chromatography. Analyst. 115: 1167-1169.

DOROUGH, H.W. (1978) Conjugation reactions of pesticides and their metabolites with sugars. In: Advances in pesticide science. Symposia papers presented at the Fourth International Congress of Pesticide Chemistry, Zurich, Switzerland, 24-28 July 1978. Part 3: Biochemistry of pests and mode of action of pesticides; pesticide degradation; pesticide residues; formulation chemistry. Oxford, New York, Pergamon Press, pp. 526-536.

DORTLAND, R.J. (1980) Toxicological evaluation of parathion and azinphosmethyl in freshwater model ecosystems. Wageningen, Centre for Agricultural Publishing and Documentation (PUDOC), 111 pp. (Agricultural Research Report No. 898).

DRAPER, W.M. & STREET, J.C. (1981) Drift from a commercial, aerial application of methyl and ethyl parathion: an estimation of human exposure. Bull. environ. Contam. Toxicol., 26: 530-536.

DU BOIS, K.P. (1961) Potentiation of the toxicity of organophosphorous compounds. Adv. Pest. Control Res., 4: 117-151.

DU BOIS, K.P. (1969) Combined effects of pesticides. Can. Med. Assoc. J., 100: 173-179.

DU BOIS, K.P. & COON, J.M. (1952) Toxicology of organic phosphorus-containing insecticides to mammals. Ind. Hyg. occup. Med., 6: 9-13.

DU BOIS, K.P. & KINOSHITA, F.K. (1968) Influence of induction of hepatic microsomal enzymes by phenobarbital on toxicity of organic phosphate insecticides (33402). Proc. Soc. Exp. Biol. Med., 129: 699-702.

DURHAM, W.F. & HAYES, W.J. Jr (1962) Organic phosphorus poisoning and its therapy. Arch. environ. Health, 6: 21-47.

EARNEST, R. (1970) Effects of pesticides on aquatic animals in the estuarine and marine environment. In: Progress in sport fishery research, 1970, Washington, DC, US Department of the Interior, Bureau of Sport Fisheries and Wildlife, pp. 10-13.

EASLEY, C.B., LAUGHLIN, J.M., GOLD, R.E., & TUPY, D.R. (1981) Methyl parathion removal from denim fabrics by selected laundry procedures. Bull. environ. Contam. Toxicol., 27: 101-108.

EBING, W. (1985) [Multimethod for the determination of pesticide residues in dead honeybees. I. Chlorine and phosphorus insecticides.] Fresenius Z. Anal. Chem., 321: 45-48 (in German).

EBING, W. (1987) [Gas chromatography of pesticides. Tabular literature abstracts, Series XV.] Berlin-Dahlem, Federal Biological Institute for Agriculture and Forestry (Report No. 236) (in German).

EICHELBERGER, J.W. & LICHTENBERG, J.J. (1971) Persistence of pesticides in river water. Environ. Sci. Technol., 5: 541-544.

EISLER, R. (1969) Acute toxicities of insecticides to marine decapod crustaceans. Crustaceana, 16(3): 300-310.

EISLER, R. (1970a) 45. Factors affecting pesticide-induced toxicity in an estuarine fish. Washington, DC, US Department of the Interior, Fish and Wildlife Service, Bureau of Sport Fisheries and Wildlife, 20 pp. (Technical Papers of the Bureau of Sport Fisheries and Wildlife No. 45/46).

EISLER, R. (1970b) 46. Acute toxicities of organochlorine and organophosphorus insecticides to estuarine fishes. Washington, DC, US Department of the Interior, Fish and Wildlife Service, Bureau of Sport Fisheries and Wildlife, 12 pp. (Technical Papers of the Bureau of Sport Fisheries and Wildlife No. 45/46).

ELKINS, E.R., FARROW, R.P., & KIM, E.S. (1972) The effect of heat processing and storage on pesticide residues in spinach and apricots. J. Agric. Food Chem., 20: 286-291.

ELLIOTT, J.W., WALKER, K.C., PENICK, A.E., & DURHAM, W.F. (1960) A sensitive procedure for urinary p-nitrophenol determination as a measure of exposure to parathion. J. Agric. Food Chem., 8: 111-113.

EMBER, M., MINDSZENTY, L., RENGEI, B., & GAL, G. (1970) Changes of vitamin A in blood and liver of organophosphorus poisoned suicides. Res. Commun. Chem. Pathol. Pharmacol., 1: 561-571.

ESPIGARES, M., ROMAN, I., GONZALEZ ALONSO, J.M., DE LUIS, B., YESTE, F., & GALVEZ, R. (1990) Proposal and application of an ecotoxicity biotest based on Escherichia coli. J. appl. Toxicol., 10: 443-446.

FABACHER, D.L. (1976) Toxicity of endrin and endrin-methyl parathion formulation to largemouth bass fingerlings. Bull Environ. Contam. Toxicol., 16: 376-378.

FAIRBROTHER, A., MEYERS, S.M., & BENNETT, R.S. (1988) Changes in mallard hen and brood behaviors in response to methyl parathion-induced illness of ducklings. Environ. Toxicol. Chem., 7: 499-503.

FAN, A.M-M. (1981) Effects of pesticides on immune competency: influence of methyl parathion and carbofuran on immunologic responses to Salmonella typhimurium infection. Diss. Abstr. Int., 41(08): 2962-B.

FAO (1985) Second Government Consultation on International Harmonization of Pesticide Registration Requirements, Rome, 11-15 October, 1982. Rome, Food and Agriculture Organization of the United Nations.

FAO/WHO (1969) 1968 Evaluation of some pesticide residues in food, Geneva, World Health Organization (FAO PL: 1968/M/9/1; WHO Food Add./69.35).

FAO/WHO (1973) 1972 Evaluation of some pesticide residues in food, Geneva, World Health Organization (AGP: 1972/M/9/1; WHO Pesticide Residues Series No. 2).

FAO/WHO (1976) 1975 Evaluation of some pesticide residues in food, Geneva, World Health Organization (AGP: 1975/M/13; WHO Pesticide Residues Series No. 5).

FAO/WHO (1980) 1979 Evaluation of some pesticide residues in food, Rome, Food and Agriculture Organization of the United Nations (FAO Plant Production and Protection Paper 20 Sup).

FAO/WHO (1981) 1980 Evaluation of some pesticide residues in food, Rome, Food and Agriculture Organization of the United Nations (FAO Plant Production and Protection Paper 26 Sup).

FAO/WHO (1985) 1984 Evaluation of some pesticide residues in food, Rome, Food and Agriculture Organization of the United Nations (FAO Plant Production and Protection Paper 67).

FAO/WHO (1986) Codex maximum limits for pesticide residues, 3rd ed. Rome, Codex Alimentarius Commission, Food and Agriculture Organization of the United Nations (CAC/Vol XIII).

FARRAN, A., CORTINA, J.L., DE PABLO, J., & BARCELO, D. (1990) On-line continuous flow extraction system in liquid chromatography with ultraviolet and mass spectrometric detection for the determination of selected organic pollutants. Anal. chim. Acta, 234: 119-126.

FAZEKAS, I.G. (1971) [Concerning the macroscopic and microscopic changes after Wofatox poisoning (methyl parathion).] Z. Rechstmed., 68: 189-194 (in German).

FAZEKAS, I.G. & RENGEI, B. (1967) [Content of methyl parathion in human organs after fatal Wofatox-poisoning.] Arch. Toxikol., 22: 381-386 (in German).

FEDORENKO, A.P., ALEEVA, L.V., & TITOK, V.A. (1981) Organophosphorus pesticide accumulation in warm-blooded animals after chemical treatment of the forest. Vestn. Zool., 4: 89-92.

FDA (1988) Food and Drug Administration pesticide program. Residues in food. Washington, DC, Food and Drug Administration, 24 pp.

FISH, S.A. (1966) Organophosphorus cholinesterase inhibitors and fetal development. J. Obstet. Gynecol., 96: 1148-1154.

FLANDERS, R.V., BLEDSOE, L.W., & EDWARDS, C.R. (1984) Effects of insecticides on *Pediobus faveolatus* (Hymenoptera: Eulophidae), a parasitoid of the Mexican bean beetle (Coleoptera: Coccinellidae). Environ. Entomol., 13: 902-906.

FLUCKE, W. (1984) [Methyl parathion - Summary toxicological information.] Wuppertal-Elberfeld, Bayer AG, Institute of Toxicology (Unpublished report submitted to WHO by Bayer AG, Leverkusen, Germany) (in German).

FLUCKE, W. & KIMMERLE, G. (1977) [Methyl parathion and ethyl parathion, studies on the acute toxicity of the two active substances.] Wuppertal-Elberfeld, Bayer AG, Institute of Toxicology (Unpublished report No. 6823, submitted to WHO by Bayer AG, Leverkusen, Germany) (in German).

FOSTER, G.D. & CROSBY, G.D. (1987) Comparative metabolism of nitroaromatic compounds in freshwater, brackish water and marine decapod crustaceans. Xenobiotica, **17**(12): 1393-1404.

FOSTER, G.D. & ROGERSON, P.F. (1990) Enhanced preconcentration of pesticides from water using the Goulden large-sample extractor. Int. J. environ. anal. Chem., **41**: 105-117.

FOSTER, R.L. (1974) Detection and measurement of ambient organophosphate pesticides. Proc. Annu. Ind. Air Pollut. Control Conf., **4**: 66-98.

FRAWLEY, J.P., WEIR, R., TUSSIG, T., DU BOIS, K.P., & CALANDRA, J.C. (1963) Toxicologic investigations on Delnav. Toxicol. appl. Pharmacol., **5**: 605-624.

FUCHS, V., GOLBE, S., KUHNERT, M., & OSSWALD, F. (1976) Studies into prenatal toxic action of parathion methyl on Wistar rats and comparison with prenatal toxicity of cyclophosphamide and trypan blue. Arch. exp. vet. Med., **30**: 343-350.

FUCHS, V., KUEHNERT, M., & GOLBS, S. (1986) Detoxifying action of humic acids on selected contaminants. Veterinärmedizin., **41**: 712-713.

FUHREMANN, T.W. (1980) Environmental behavior of insecticides. Diss. Abstr. Int. B., **40**(7): 3075-3076.

FUHREMANN, T.W. & LICHTENSTEIN, E.P. (1978) Release of soil-bound methyl (^{14}C)parathion residues and their uptake by earthworms and oat plants. J. Agric. Food Chem., **26**(3): 605-610.

FUKAMI, J. & SHIHIDO, T. (1963) Studies on the selective toxicities of organic phosphorous insecticides (III). Botyu Kagaku, **28**: 77-81.

FUNCH, F.H. (1981) Analysis of residues of seven pesticides in some fruits and vegetables by high-pressure liquid chromatography. Z. Lebensm.-Unters. Forsch, **173**: 95-98.

GABICA, J., WYLLIE, J., WATSON, M., & BENSON, W.W. (1971) Example of flame photometric analysis for methyl parathion in rat whole blood and brain tissue. Anal. Chem., **43**: 1102-1105.

GAINES, T.B. (1960) The acute toxicity of pesticides to rats. Toxicol. appl. Pharmacol., **2**: 88-99.

GAINES, T.B. (1969) Acute toxicity of pesticides. Toxicol. appl. Pharmacol., **14**: 515-534.

GAJAN, R.J. (1969) Collaborative study of confirmative procedures by single-sweep oscillographic polarography for the determination of organophosphorus pesticide residues in nonfatty foods. J. Assoc. Off. Anal. Chem., **52**: 811-817.

GAMBRELL, R.P., TAYLOR, B.A., REDDY, K.S., & PATRICK, W.H. Jr (1984) Fate of selected toxic compounds under controlled redox potential and pH conditions in soil and sediment-water systems. Athens, Georgia, US Environmental Protection Agency, Office of Research and Development, Environmental Research Laboratory.

GAR, K.A., SAZONOVA, N.A., FADEEV, Y.N., VLADIMIROVA, I.L., & GOLUBEVA, Z.Z. (1958) [Incorporation and excretion of dimethyl 4-nitrophenyl thiophosphate in guinea pigs.] Org. Insektofungits. Gerbits., **1958**: 93-105 (in Russian).

GARRIDO, S.J.J. & MONTEOLIVA, H.M. (1981) [Qualitative and quantitative determination of residues of diazinon, dimethoate, ethion, malathion, methyl parathion and parathion in soil extracts by thin-layer chromatography, and confirmation with gas chromatography.] An. Edafol. Agrobiol., **40**: 1787-1798 (in Spanish).

GAUGHAN, L.C., ENGEL, J.L., & CASIDA, J.E. (1980) Pesticide interactions: effects of organophosphorus pesticides on the metabolism, toxicity, and persistence of selected pyrethroid insecticides. Pestic. Biochem. Physiol., **14**: 81-85.

GERSTL, Z. & HELLING, C.S. (1985) Fate of bound methyl parathion residues in soils as affected by agronomic practices. Soil Biol. Biochem., **17**(5): 667-673.

GHOSH, P., BHATTACHARYA, S., & BHATTACHARYA, S. (1989) Impact of nonlethal levels of metacid-50 and carbaryl on thyroid function and cholinergic system of *Channa punctatus*. Biomed. environ. Sci., **2**: 92-97.

GHOSH, P., BHATTACHARYA, S., & BHATTACHARYA, S. (1990) Impairment of the regulation of gonadal function in *Channa punctatus* by metacid-50 and carbaryl under laboratory and field conditions. Biomed. environ. Sci., **3**: 106-112.

GILE, J.D. & GILLETT, J.W. (1981) Transport and fate of organophosphate insecticides in a laboratory model. J. Agric. Food Chem., **29**: 616-621.

GILLESPIE, A.M. & WALTERS, S.M. (1986) HPLC silica column fractionation of pesticides and PCB from butterfat. J. liq. Chromatogr., **9**: 2111-2141.

GILLESPIE, A.M. & WALTERS, S.M. (1989) Semi-preparative reverse phase HLPC fractionation of pesticides from edible fats and oils. J. liq. Chromatogr., **12**(9): 1687-1703.

GILLETT, J.W. & GILE, J.D. (1976) Pesticide fate in terrestrial laboratory ecosystems. Int. J. environ. Stud., **10**: 15-22.

GILOT-DELHALLE, J., COLIZZI, A., MOUTSCHEN, J., & MOUTSCHEN-DAHMEN, M. (1983) Mutagenicity of some organophosphorus compounds at the *ade*6 locus of *Schizosaccharomyces pombe*. Mutat. Res., **117**: 139-148.

GOEDICKE, H.-J. (1989) [Exposure to residues on leaf surfaces following the use of organophosphorus insecticides in high-intensity apple growing.] Z. Gesamte Hyg., **35**: 533-535 (in German).

GOEDICKE, H.-J. & WINKLER, R. (1976) [On the residue of methyl parathion in soil.] Nachrichtenbl. Pflanzenschutz., **30**(5): 100-101 (in German).

GOLOVATYI, V.G., KOROL, E.N., KLISENKO, M.A., & GIRENKO, D.B. (1982) FDMS study of pesticides. Eur. J. Mass Spectrom. biochem., Med. environ. Res., 2: 97-100.

GONCHARUK, E.I., LIPATOVA, T.E., & FILATOVA, I.N. (1988) Conditions for elevation of pesticide content in the atmospheric boundary layer of agricultural fields. Gig. i Sanit., 1: 25-27.

GOYER, R. & CHEYMOL, J. (1967) Contribution à l'étude pharmacologique de quelques dérivés soufrés de l'acide orthophosphorique trisubstitué. II. Essais de toxicité. Rev. Can. Biol., 26: 35-41.

GOZA, S.W. (1972) Infrared analysis of pesticide formulations. J. Assoc. Off. Anal. Chem., 55: 913-917.

GREENHALGH, R., BLACKWELL, B.A., PRESON, C.M., & MURRAY, W.J. (1983) Phosphorus-31 nuclear magnetic resonance analysis of technical organophosphorus insecticides for toxic contaminants. J. Agric. Food Chem., 31: 710-713.

GRETCH, F.M. & ROSEN, J.D. (1984) Automated sample cleanup for pesticide multiresidue analysis. Part II. Design and evaluation of column chromatographs module. J. Assoc. Off. Anal. Chem., 67: 783-789.

GRETCH, F.M. & ROSEN, J.D. (1987) Automated sample cleanup for pesticide multiresidue analysis. Part III. Evaluation of complete system for screening subtolerance residues in vegetables. J. Assoc. Off. Anal. Chem., 70: 109-111.

GROVER, I.S. & MALHI, P.K. (1985) Genotoxic effects of some organophosphorus insecticides. I. Induction of micronuclei in bone marrow cells in rats. Mutat. Res., 155: 131-134.

GROVER, I.S. & MALHI, P.K. (1987) Genotoxic effects of some organophosphorus pesticides. II. *In vivo* chromosomal aberration bioassay in bone marrow cells in rat. Mutat. Res., 188: 45-51.

GRUE, C.E. (1982) Response of common grackles to dietary concentrations of four organophosphate pesticides. Arch. environ. Contam. Toxicol., 11: 617-626.

GUPTA, R.C. & KADEL, W.L. (1990) Methyl parathion acute toxicity: prophylaxis and therapy with memantine and atropine. Arch. int. Pharmacodyn., 305: 208-221.

GUPTA, R.C., THORNBURG, J.F., STEDMAN, D.B., & WELSCH, F. (1984) Effect of subchronic administration of methyl parathion on *in vivo* protein synthesis in pregnant rats and their conceptuses[1,2]. Toxicol. appl. Pharmacol., 92: 457-468.

GUPTA, R.C., RECH, R.H., LOVELL, K.L., WELSCH, F., & THORNBURG, J.E. (1985) Brain cholinergic, behavioral, and morphological development in rats exposed *in utero* to methylparathion. Toxicol. appl. Pharmacol., 77: 405-413.

GYORFI, L., AMBRUS, A., & BOLYGO, E. (1987) Optimization of determination and clean-up parameters for sensitive multiresidue analysis of pesticides. In: Greenhalgh, P. & Roberts, T.R., ed. IUPAC. Pesticide science and biotechnology. Proceedings of the Sixth International Congress on Pesticide Chemistry, Ottawa, Canada, 10-15 August 1986. Oxford, London, Blackwell Scientific Publications, pp. 353-356.

HALEY, T.J., FARMER, J.H., HARMON, J.R., & DOOLEY, K.L. (1975) Estimation of the LD_1 and extrapolation of the $LD_{0.1}$ for five organothiophosphate pesticides. Eur. J. Toxicol., **8**: 229-235.

HANSCH, C. & LEO, A. (1987) The log p database. Claremont, California, Pomona College, 268 pp.

HAPKE, H.-J., YOUSSEF, S., & OMER, O.H. (1978) [Modification of the alkylphosphate toxicity in lead contaminated rats.] Dtsch. Tierärztl. Wochenschr., **85**: 321-456 (in German).

HAQUE, R. & FREED, V.H. (1974) Behavior of pesticides in the environment: Environmental chemodynamics. Residue Rev., **52**: 89-116.

HASAN, M. & AHMAD KHAN, N. (1985) Methyl parathion-induced dose-related alteration in lipid levels and lipid peroxidation in various regions of rat brain and spinal cord. Ind. J. exp. Biol., **23**: 141-144.

HATCHER, J.S. & WISEMAN, R.L. (1969) Epidemiology of pesticide poisoning in the lower Rio Grande Valley in 1968. Texas Med., **65**: 40-43.

HAYES, W.J. Jr (1971) Studies on exposure during the use of anticholinesterase pesticides. Bull. World Health Organ., **44**: 272-288.

HAYES, W.J. Jr & LAWS, E.R. Jr (1991) Handbook of pesticide toxicology. New York, London, San Francisco, Academic Press, vol. 1-3.

HECHT, G. & WIRTH, W. (1950) [The pharmacology of the phosphoric acid ester derivatives of thiophoric acid.] Arch. exp. Pharmakol., **211**: 264-277 (in German).

HEIMANN, K.G. (1982) [E 120 (methyl parathion):studies of acute oral and acute dermal toxicity.] Wuppertal-Elberfeld, Bayer AG, Institute of Toxicology (Unpublished report No. 10897, submitted to WHO by Bayer AG, Leverkusen, Germany) (in German).

HENNIGHAUSEN, G. (1984) [On the toxicological significance of the interaction of foreign compounds with glutathione and glutathione S-transferases.] Z. Gesamle Hyg., **30**(11): 603-606 (in German with English summary).

HERBERT, G.B., PETERLE, T.J., & GRUBB, T.C. (1989) Chronic dose effects of methyl parathion on nuthatches: cholinesterase and ptilochronology. Bull. environ. Contam. Toxicol., **42**: 471-475.

HERBOLD, B. (1986) [E 120, C.N. methyl parathion-salmonella/microsome test to evaluate for point mutagenic effect.] Wuppertal-Elberfeld, Bayer AG, Institute of Toxicology (Unpublished report No. 15306, submitted to WHO by Bayer AG, Leverkusen, Germany) (in German).

HICKE, K. & THIEMANN, W. (1987) [The decomposition of selected phosphoric acid esters by UV-irradiation.] Vom Wasser, **69**: 85-94 (in German).

HILL, E.F. & CAMARDESE, M.B. (1986) Lethal dietary toxicities of environmental contaminants and pesticides to *Coturnix*. Washington, DC, US Department of the Interior Fish and Wildlife Service, pp. 100-101 (Fish and Wildlife Technical Report No. 2).

HILL, E.F., HEATH, R.G., SPANN, J.W., & WILLIAMS, J.D. (1975) Lethal dietary toxicities of environmental pollutants to birds. Washington, DC, US Department of the Interior Fish and Wildlife Service, pp. 16, 27 (Fish and Wildlife Technical Report No. 191).

HIRAYAMA, K. & TAMAOI, S. (1980) Acute toxicity of methylparathion and diazinon (pesticide) to larvae of kuruma praure *Penaeus japonicus* and of swimming crab *Portunus trityberculatus*. Bull. Jpn Soc. Sci. Fish., **46**: 117-123.

HIRSCHELMANN, R. & BEKEMEIER, H. (1975) Acute toxicity of demephion and parathion methyl in dogs. In: Proceedings of the European Society of Toxicology. Amsterdam Excerpta Medica Foundation, pp. 273-275.

HOLLINGWORTH, R.M., METCALF, R.L., & FUKUTO, T.R. (1967) The selectivity of sumithion compared with methyl parathion. Metabolism in the white mouse. J. Agric. Food Chem., **15**: 242-249.

HOLM, H.W., KOLLIG, H.P., & PAYNE, W.R. Jr (1983) Fate of methyl parathion in aquatic channel mirocosms. Environ. Toxicol. Chem., **2**: 169-176.

HOLMSTEAD, R.L. & CASIDA, J.E. (1974) Chemical ionization mass spectrometry of organophosphorus insecticides. J. Assoc. Off. Anal. Chem., **57**: 1050-1055.

HUANG, C.C. (1973) Effect on growth but not on chromosomes of the mammalian cells after treatment with three organophosphorus insecticides. Proc. Soc. Exp. Biol. Med., **142**: 36-40.

HUDSON, R.H., HAEGELE, M.A., & TUCKER, R.K. (1979) Acute oral and percutaneous toxicity of pesticides to mallards: correlations with mammalian toxicity data. Toxicol. appl. Pharmacol., **47**: 451-460.

HUDSON, R.H., TUCKER, R.K., & HAEGELE, M.A. (1984) Handbook of toxicity of pesticides to wildlife, 2nd ed. Washington, DC, US Department of the Interior, Fish and Wildlife Service, pp. 52-53 (Fish and Wildlife Publication, No. 153).

HUMMEL, S.V. & YOST, R.A. (1986) Tandem mass spectrometry of organophosphate and carbamate pesticides. Org. mass Spectrom., **21**: 785-791.

IARC (1983) Methyl parathion. In: Miscellaneous pesticides. Lyon, International Agency for Research on Cancer, pp. 131-152 (IARC Monographs on the evaluation of the carcinogenic risk of chemicals to humans, Volume 30).

IARC (1987) Methyl parathion. In: Overall evaluations of carcinogenicity: an updating of IARC monographs, Volumes 2 to 42. Lyon, International Agency for Research on Cancer, pp. 392 (IARC Monographs on the evaluation of the carcinogenic risk of chemicals to humans, Supplement 7).

IZMIROVA, N., SHALASH, S., & KALOYANOVA, F. (1984) [Dynamics of cholinesterase inhibition in methylparathion poisoning.] Probl. Khig., **9**: 42-49 (in Bulgarian).

JACKSON, E.R. (1976) High-performance and gas-liquid chromatography of methyl parathion formulations. J. Assoc. Off. Anal. Chem., **59**: 740-744.

JACKSON, E.R. (1977a) Collaborative study of gas-liquid chromatographic method for determining methyl parathion. J. Assoc. Off Anal. Chem., **60**: 720-723.

JACKSON, E.R. (1977b) Collaborative study of high pressure liquid chromatographic method for determining methyl parathion. J. Assoc. Off. Anal. Chem., **60**: 724-727.

JACKSON, M.D. & LEWIS, R.G. (1978) Volatilization of two methyl parathion formulations from treated fields. Bull. environ. Contam. Toxicol., **20**: 793-796.

JAGLAN, P.S. & GUNTHER, F.A. (1970) Comparison of hydrolysis rates of methyl parathion and methyl paraoxon by gas liquid chromatography and spectrometry. J. chromatogr. Sci., **8**: 483-485.

JALALUDDIN, M. & MOHANASUNDARAM, M. (1989) Residual toxicity of four insecticides recommended for control of coconut coccids on the parasitoid fauna of *Opisina arenosella* WLK. Entomol., **14**: 199-202.

JARVINEN, A.W. & TANNER, D.K. (1982) Effects of selected controlled release and corresponding unformulated technical grade pesticides to the fathead minnow *Pimephales promelas*. Environ. Pollut., Ser. A, **27**(3): 179-195.

JELINEK, R., PETERKA, M., & RYCHTER, Z. (1985) Chick embryotoxicity screening test - 130 substances tested. Indian J. exp. Biol., **23**: 588-595.

JOHANSSON, C.E. (1978) A multiresidue analytical method for determining organochlorine, organophosphorus, dinitrophenyl and carbamate pesticides in apples. Pestic. Sci., **9**: 313-322.

JOHNSON, R.D. & FINLEY, M.T. (1980) Handbook of acute toxicity of chemicals to fish and aquatic invertebrates. Summaries of toxicity tests conducted at Columbia National Fisheries Research Laboratory, 1965-78. Washington, DC, US Department of the Interior Fish and Wildlife Service, 98 pp. (Resource Publication 137).

JOHNSON, R.D. & MANSKE, D.D. (1976) Residues in food and feed. Pesticide residues in total diet samples (IX). Pestic. Monit. J., **9**: 157-169.

JOHNSON, R.D., MANSKE, D.D., NEW, D.H., & PODREBARAC, D.S. (1981) Food and feed - pesticide, heavy metal, and other chemical residues in infant and toddler. Pestic. Monit. J., **15**: 39-50.

JOHNSON, M.W., WELTER, S.C., TOSCANO, N.C., & IWATA, Y. (1983) Lettuce yield reductions correlated with methyl parathion use. J. Econ. Entomol., **76**: 1390-1394.

JONES, D.C.L., SIMMON, V.F., MORTELMANS, K.E., MITCHELL, A.D., EVANS, E.L., JOTZ, M.M., RICCIO, E.S., ROBINSON, D.E., & KIRKHART, B.A. (1982) In vitro mutagenicity studies of environmental chemicals. Research Triangle Park, North Carolina, US Environmental Protection Agency, Office of Research and Development, Health Effects Research Laboratory (EPA-600/1-84-003; PB84-138973).

JOSHI, U.M. & THORNBURG, J.E. (1986) Interactions between cimetidine, methylparathion, and parathion. J. Toxicol. environ. Health, **19**: 337-344.

KAGAN, J.S. (1971) [Topical questions about the toxicology of phosphororganic insecticides.] Ernährungsforschung, **14**: 503-514 (in German).

KALOYANOVA-SIMEONOVA, F.P. (1970) Prevention of intoxication by pesticides in Bulgaria. In: Proceedings of the Fourth International Congress on Rural Medicine, Usuda (Tokyo), 1969. Tokyo, Japanese Association of Rural Medicine, pp. 35-37.

KAMIENSKI, F.X. & MURPHY, S.D. (1971) Biphasic effects of methylenedioxyphenyl synergists on the action of hexobarbital and organophosphate insecticides in mice. Toxicol. appl. Pharmacol., 18: 883-894.

KARCSU, S., TOTH, L., MAROSI, G., & SIMON, Z. (1981) Analysis of histochemical changes due to acute wofatox poisoning in a model experiment. Morphol. Igazsagugyi Orv. Sz., 21: 289-296.

KARLOG, O., NIELSEN, P., & RASMUSSEN, F. (1978) Toxicokinetics. Arch. Toxicol., 1(Suppl.): 55-67.

KATKAR, H.N. & BARVE, V.P. (1976) A new spray reagent for the identification and determination of organophosphorus insecticides by thin layer chromatography. Curr. Sci., 45: 662-664.

KAWAHARA, F.K., LICHTENBERG, J.J., & EICHELBERGER, J.W. (1967) Thin-layer and gas chromatographic analysis of parathion and methyl parathion in the presence of chlorinated hydrocarbons. J. Water Pollut. Control Fed., 39: 446-457.

KAZAKOVA, M.V., YAKUB, G.G., & MANDRIK, F.I. (1974) Quantitative detection of metaphos (methyl parathion) in feeds and its effect on animals. Veterinariya (Moscow), 2: 107.

KHAN, M.S. (1988) An electrolytic reduction method for the determination of fenitrothion and methyl paration. Pak. J. Sci. Ind. Res., 31: 20-22.

KHAN, M.S. & HASAN, M. (1988) Dose-related neurochemical changes in the levels of gangliosides and glycogen in various regions of the rat brain and spinal cord following methyl parathion administration. Exp. Pathol., 35: 61-65.

KHATTAT, F.H. & FARLEY, S. (1976) Acute toxicity of certain pesticides to *Acartia tonsa* Dana. Narangansett, Rhode Island, US Environmental Protection Agency, Office of Research and Development, Environmental Research Laboratory, 23 pp. (EPA-600/3-76-033).

KHEIFETS, L.Y., SOBINA, N.A., & ROMANOV, N.A. (1976) [Use of differential oscillopolarography for determining substances with limitable concentrations in water.] Probl. Okhr. Vod, 7: 21-24 (in Russian).

KHEIFETS, L.Y., ROMANOV, N.A., & SOBINA, N.A. (1980) [Analytical possibility of differential chronoamperometry with dropping electrodes.] Zh. Anal. Khim., 35: 874-879 (in Russian).

KIDO, H., BAILEY, J.B., MCCALLEY, N.F., YATES, W.E., & COWDEN, R.E. (1975) The effect of overhead sprinkler irrigation on methyl parathion residue on grape leaves. Bull. environ. Contam. Toxicol., 14: 209-213.

KIMMERLE, G. (1975) [Testing of DDT for combined effect with methyl parathion and fenitrothion in acute tests on rats.] Wuppertal-Elberfeld, Bayer AG, Institute of Toxicology (Unpublished report No. 5591, submitted to WHO by Bayer AG, Leverkusen, Germany) (in German).

KIMMERLE, G. & LORKE, D. (1968) [Toxicology of insecticidal phosphoric acid esters.] Pflanzenschutz.-Nachr. BAYER, **21**: 111-142 (in German).

KJOELHOLT, J. (1985) Determination of trace amounts of organophosphorus pesticides and related compounds in soils and sediments using capillary gas chromatography and a nitrogen-phosphorus detector. J. Chromatogr., **325**: 231- 238.

KLISENKO, M.A. & GIRENKO, D.B. (1980) [Gas-chromatographic determination of organophosphorus pesticides in air.] Gig. i Sanit. **1980**(9): 57-60 (in Russian).

KLISENKO, M.A., GIRENKO, D.B., GOLOVATYI, V.G., SHPAKOVSKII, I.V., & KOROL, E.N. (1981) [Field-desorption mass spectrometry in pesticide analysis.] Khim. Sel'sk. Khoz. **1981**: 59-61 (in Russian).

KNIEHASE, U. & ZOEBELEIN, G. (1990) [Testing the effects of pesticides on the predator mite *Phytoseiulus persimilis* Ath.-Hen. by means of a new laboratory method approaching to the practice.] Anz. Schädlingskd. Pflanzenschutz. Umweltschutz., **63**: 105-113 (in German).

KOEN, J.G. & HUBER, J.F.K. (1970) A rapid method for residue analysis by column liquid chromatography with polarographic detection. Application to the determination of parathion and methylparathion on crops. Anal. chim. Acta, **51**: 303-307.

KONRAD, J.G., PIONKE, H.B., & CHESTERS, G. (1969) Extraction of organochlorine and organophosphate insecticides from lake waters. Analyst, **94**: 490-492.

KORN, S. & EARNEST, R. (1974) Acute toxicity of twenty insecticides to striped bass, *Morone saxatilis*. Calif. Fish Game, **60**(3): 128-131.

KORSOS, I. & LANTOS, J. (1984) [Thin-layer chromatographic identification of pesticides.] Novenyvedelem (Budapest), **20**: 30-34 (in Hungarian).

KRIJGSMAN, W. & VAN DE KAMP, C.G. (1976) Analysis of organophosphorus pesticides by capillary gas chromatography with flame photometric detection. J. Chromatogr., **117**: 201-205.

KRUEGER, H.R. & CASIDA, J.E. (1957) Toxicity of fifteen organophosphorus insecticides to several insect species and to rats. J. Econ. Entomol., **50**: 356-358.

KUMAR, N.V.N. (1985) Chromatographic and enzymic methods as applied to environmental monitoring of pesticide residues in some districts of Andhra Pradesh. In: Trivedy, R.K. & Goel, P.K. ed. Current pollution research in India. Kared, India, Environmental Publications, pp. 163-184.

KUMAR, N.V.N. & RAMASUNDARI, M. (1980) Colorimetric determination of methyl parathion and oxygen analog. J. Assoc. Off. Anal. Chem., **63**: 536-538.

KUMMER R. & VAN SITTERT, N.J. (1986) Field studies on health effects from the application of two organophosphorus insecticide formulations by hand-held ULV to cotton. Toxicol. Lett., **33**: 7-24.

LAMONTAGNE, J.C. (1978) Résidues de produits phytosanitaires dans les fruits et légumes. C. R. Séances Acad, Agric. France, **64**: 923-931.

LANGE, P. & WIEZOREK, W.D. (1975) Effects of diethyldithiocarbamate on acute toxicity and acetylcholinesterase inhibition by methylparathion in mice. Acta biol. med. Germ., **34**: 427-433.

LANGE, P., TIEFENBACH, B., & WIEZOREK, W.D. (1977) Effect of dithiocarb on acute toxicity and acetylcholinesterase inhibition by different organophosphates in mice. Proc. Eur. Soc. Toxicol., **18**: 284-285.

LASSITER, R.R., PARRISH, R.S., & BURNS, L.A. (1986) Decomposition by planktonic and attached microorganisms improves chemical fate models. Environ. Toxicol. Chem., **5**: 29-39.

LAUGHLIN, J. & GOLD, R.E. (1989) Evaporative dissipation of methyl parathion from laundered protective apparel fabrics. Bull. environ. Contam. Toxicol., **42**: 566-573.

LAUGHLIN, J.M., EASLEY, C.B., GOLD, R.E., & TUPY, D.R. (1981) Methyl parathion transfer from contaminated fabrics to subsequent laundry and to laundry equipment. Bull. environ. Contam. Toxicol., **27**: 518-523.

LAWRENCE, J.F. & TURTON, D. (1978) High performance liquid chromatographic data to 166 pesticides. J. Chromatogr., **159**: 207-226.

LAWRENCE, J.F., RENAULT, C., & FREI, R.W. (1976) Fluorogenic labelling of organophosphate pesticides with dansyl chloride. Application to residue analysis by high-pressure liquid chromatography and thin-layer chromatography. J. Chromatogr., **121**: 343-351.

LEARD, R.L., GRANTHAM, B.J., & PESSONEY, G.F. (1980) Use of selected freshwater bivalves for monitoring organochlorine pesticide residues in major Mississippi stream systems, 1972-1973. Pestic. Monit. J., **14**: 47-52.

LE BEL, G.L., WILLIAMS, D.T., GRIFFITH, G., & BENOIT, F.M. (1979) Isolation and concentration of organophosphorus pesticides from drinking water at the ng/L level, using macroreticular resin. J. Assoc. Off. Anal. Chem., **62**: 241-249.

LEE, H.B., WENG, L.D., & CHAU, A.S.Y. (1984) Confirmation of pesticide residue identity. XI. Organophosphorus pesticides. J. Assoc. Off. Anal. Chem., **67**: 553-556.

LEIDY, R.B., SHEETS, T.J., & NELSON, L.A. (1977) Residues from two formulations of methyl parathion on flue-cured tobacco. Tob. Sci., **21**: 28-31.

LESHCHEV, V.V. & TALANOV, G.A. (1977) [Determination of residues of organophosphorus pesticides using thin-layer chromatography with enzymic development.] Khim. Sel'sk. Khoz., **15**: 46-48 (in Russian).

LEVINE, B.S. & MURPHY S.D. (1977a) Esterase inhibition and reactivation in relation to piperonyl butoxide-phosphorothionate interactions. Toxicol. appl. Pharmacol., **40**: 379-391.

LEVINE, B.S. & MURPHY, S.D. (1977b) Effect of piperonyl butoxide on the metabolism of dimethyl and diethyl phosphorothionate insecticides. Toxicol. appl. Pharmacol., **40**: 393-406.

LEWIS, D.L. & HOLM, H.W. (1981) Rates of transformation of methyl parathion and diethyl phthalate by aufwuchs microorganisms. Appl. environ. Microbiol., **42**: 698-703.

LEWIS, D.L., HODSON, R.E., & FREEMAN, L.F. III (1984) Effects of microbial community interactions on transformation rates of xenobiotic chemicals. Appl. environ. Microbiol., **48**: 561-565.

LEWIS, D.L., HODSON, R.E., & FREEMAN, L.F. III (1985) Multiphasic kinetics for transformation of methyl parathion by *Flavobacterium* species. Appl. environ. Microbiol., **50**: 553-557.

LEWIS, R.G. & JACKSON, M.D. (1982) Modification and evaluation of a high-volume air sampler for pesticides and semiolatile industrial organic chemicals. Anal. Chem., **54**: 592-594.

LEWIS, R.G. & MACLEOD, K.E. (1982) Portable sampler for pesticides and semivolatile industrial organic chemicals in air. Anal. Chem., **54**: 310-315.

LEWIS, R.G., BROWN, A.R., & JACKSON, M.D. (1977) Evaluation of polyurethane foam for sampling of pesticides, polychorinated biphenyls and polychlorinated naphthalenes, in ambient air. Anal. Chem., **49**: 1668-1672.

LI, Y. & WANG, G. (1987) [Determination of organophosphorus pesticides in wastewater.] Shanghai Huanjing Kexue, **6**: 19-31, 42 (in Chinese).

LIANG, L. & ZHANG, G. (1986) [Gas chromatographic determination of organophosphorus agrochemicals in the air.] Zhonghua Laodong Weisheng Zhiyebing Zazhi, **4**: 163-165 (in Chinese).

LICHTENSTEIN, E.P. (1975) Transport mechanism in soil. Metabolism and movement of insecticides from soils into water and crop plants. Pure appl. Chem., **42**(1/2): 113-118.

LICHTENSTEIN, E.P. & SCHULZ, K.R. (1964) The effects of moisture and microorganisms on the persistence and metabolism of some organophosphorous insecticides in soils, with special emphasis on parathion. J. Econ. Entomol., **57**: 618-627.

LISI, P., CARAFFINI, S., & ASSALVE, D. (1986) A test series for pesticide dermatitis. Contact Dermatitis, **15**: 266-269.

LISI, P., CARAFFINI, S., & ASSALVE, D. (1987) Irritation and sensitization potential of pesticides. Contact Dermatitis, **17**: 212-218.

LITTLE, E.E., ARCHESKI, R.D., FLEROV, B.A., & KOZLOVSKAYA, V.I. (1990) Behavioural indicators of sublethal toxicity in rainbow trout. Arch. environ. Contam. Toxicol., **19**: 380-385.

LOFROTH, G., KIM, C.H., & HUSSAIN, S. (1969) Alkylating properties of 2,2-dichlorovinyl dimethyl phosphate: a disregarded hazard. Environ. Mutat. Soc. Newsl., **2**: 21-27.

LOSER, E. & EIBEN, R. (1982) [E 605-methyl: Multigeneration tests on rats.] Wuppertal-Elberfeld, Bayer AG, Institute of Toxicology (Unpublished report No. 10630, submitted to WHO by Bayer AG, Leverkusen, Germany) (in German).

LOVE, J.L., DONNELLY, R.G.C., HUGHES, J.T., & WILSON, P.D. (1974) Pesticide residues in fruit and vegetables in New Zealand 1971-73. N. Z. J. Sci., **17**: 529-534.

LUKE, M.A. & DOOSE, G.M. (1983) A modification of the Luke multiresidue procedure for low moisture, nonfatty products. Bull. environ. Contam. Toxicol., **30**: 110-116.

LUKE, M.A. & DOOSE, G.M. (1984) A rapid analysis for pesticides in milk and oilseeds. Bull. environ. Contam. Toxicol., **32**: 651-656.

LUKE, M.A., FROBERG, J.E., & MASUMOTO, H.T. (1975) Extraction and cleanup of organochlorine, organophosphate, organonitrogen, and hydrocarbon pesticides in produce for determination by gas-liquid chromatography. J. Assoc. Off. Anal. Chem., **58**: 1020-1026.

McCANN, J.A. & JASPER, R.L. (1972) Vertebral damage to bluegills exposed to acutely toxic levels of pesticides. Trans. Am. Fish. Soc., **101**(2): 317-322.

McCOLLISTER, D.D., OLSON, K.J., ROWE, V.K., PAYNTER, O.E., WEIR, R.J., & DIETERICH, W.H. (1968) Toxicology of 4-tert-butyl-2-chlorophenyl methyl methylphosphoramidate (ruelene) in laboratory animals. Food Cosmet. Toxicol., **6**: 185-198.

MACEK, K.J. & McALLISTER, W.A. (1970) Insecticide susceptibility of some common fish family representatives. Trans. Am. Fish Soc., **99**(1): 20-27.

MACHEMER, L. (1977a) [Methyl parathion. Embryotoxic or teratogenic effects in rats after oral application.] Leverkusen, Germany, Bayer AG, Institute for Toxicology, (Unpublished report submitted to WHO by Bayer AG) (in German).

MACHEMER, L. (1977b) [Methyl parathion. Tests for embryotoxic and teratogenic effects in rats after intravenous injection.] Wuppertal-Elberfeld, Bayer AG, Institute of Toxicology (Unpublished report No. 6812, submitted to WHO by Bayer AG, Leverkusen, Germany) (in German).

McLEOD, K.E & LEWIS, R.G. (1982) Portable sampler for pesticides and semivolatile industrial organic chemicals in air. Anal. Chem., **54**: 310-315.

MAFF (1986) Report of the Working Party on Pesticide Residues (1982-85). The sixteenth report of the Steering Group on Food Surveillance. London, Ministry of Agriculture, Fisheries and Food, Health and Safety Executive, 50 pp. (Food Surveillance Paper No. 16).

MAFF (1990) Report of the Working Party on Pesticide Residues (1988-89). (Supplement to issue No. 8 (1990) of the Pesticide Register). London, Ministry of Agriculture, Fisheries and Food, Health and Safety Executive, 86 pp.

MANSKE, D.D. & JOHNSON, R.D. (1975) Residues in food and feed. Pesticide residues in total diet samples (VIII). Pestic. Monit. J., **9**: 94-105.

MANSKE, D.D. & JOHNSON, R.D. (1977) Pesticide and other chemical residues in total diet samples (X). Pestic. Monit. J., **10**: 134-148.

MARCUS, M., SPIGARELLI, J., & MILLER, H. (1978) Organic compounds in organophosporus pesticide manufactoring wastewaters. Springfield, Virginia, National Technical Information Service, p. 22 (PB-289821).

MARTIN, E.W. (1978) Hazards of medication. A manual on drug interactions, contraindications, and adverse reactions with other presribing and drug information, 2nd ed. Philadelphia, Toronto, J.B. Lippincott Company.

MATHEW, G., ABDUL RAHIMAN, M., & VIJAYALAXMI, K.K. (1990) *In vivo* genotoxic effects in mice of Metacid 50 an organophosphorus insecticide. Mutagenesis, 5(2): 147-149.

MAYER, F.L. Jr (1987) Acute toxicity handbook of chemicals to estuarine organisms. Gulf Breeze, Florida, US Environmental Protection Agency, 283 pp.

MAYER, F.L. Jr & ELLERSIECK, M.R. (1986) Manual of acute toxicity: interpretation and data base for 410 chemicals and 66 species of freshwater animals. Washington, DC, US Department of the Interior Fish and Wildlife Service, pp. 311 (Resource Publication 160).

MELNIKOW, N.N. (1971) Chemistry of pesticides. Berlin, Heidelberg, New York, Springer Verlag.

MENZIE, C.M. (1972) Fate of pesticides in the environment. Ann. Rev. Entomol., 17: 199-222.

MENZIE, C.M. (1974) Metabolism of pesticides - an update. Washington, DC, US Department of the Interior, Fish and Wildlife Service (Report No. 184).

MESTRES, R., LEONARDI, G., CHEVALLIER, C., & TOURTE, J. (1969) Résidus de pesticides. XIX. Méthode de recherche des résidus de pesticides dans les eaux naturelles. 1ère partie. Méthode d'analyse générale. Ann. Fals. Expert. Chim., 62(685): 75-85.

MESTRES, R., CHEVALLIER, C., ESPINOZA, C., & CORNET, R. (1977) XXXIII. Application du couplage chromatographie gazeuse spectrométrie de masse aux méthodes de recherche et de dosage des résidus de pesticides et de micropolluants organiques dans les matéiaux de l'environnement et les matières alimentaires. Ann. Fals. Expert. Chim., 70: 177-188.

MEYER JONES, L., BOOTH, N.H. & McDONALD, L.E. (1977) Veterinary pharmacology and therapeutics, 4th ed. Ames, Iowa State University Press, pp. 1201.

MEYERS, S.M., CUMMINGS, J.L., & BENNETT, R.S. (1990) Effects of methyl parathion on red-winged blackbird (*Agelaius phoeniceus*) incubation behavior and nesting success. Environ. Toxicol. Chem., 9: 807-813.

MIDWEST RESEARCH INSTITUTE (1975) Substitute chemical program - initial scientific and minieconomic review of methyl parathion. Kansas City, Missouri, Midwest Research Institute, 176 pp. (Prepared for US Environmental Protection Agency, Criteria and Evaluation Division, Washington, DC.) (EPA-540/1-75-004).

MIELLET, A. (1982) Dosage des résidus de pesticides par C.L.H.P. sur plantes médicinales et aromatiques. Compte-rendu du résultat d'applications de pesticides effectuées avec le concours du syndicat national des plantes médicinales de Nilly-la-Forét. Ann. Fals. Expert. Chim., 75: 369-375.

MIHAIL, F. & VOGEL, O. (1984) [E 120/methyl parathion: Subacute dermal toxicity tests on rabbits.] Wuppertal-Elberfeld, Bayer AG, Institute of toxicology (Unpublished report No. 12484, submitted to WHO by Bayer AG, Leverkusen, Germany) (in German).

MILES, J.W., FETZER, L.E., & PEARCE, G.W. (1970) Collection and determination of trace quantities of pesticides in air. Environ. Sci. Technol., 4(5):420-425.

MILLER, H., CRAMER, P., DRINKWINE, A., SHAN, A., RISCHAN, G., & GOING, J. (1981) Development of methods for pesticides in wastewater: applicability of a general approach. In: Keith, L.H., ed. Advances in the identification and analysis of organic pollutants in water. Ann Arbor, Michigan, Ann Arbor Science Publishers, Vol. 1, pp. 115-138.

MILLS, P.A., ONLEY, J.H., & GAITHER, R.A. (1963) Rapid method for chlorinated pesticide residues in vegetables and food. J. Am. Off. Anal. Chem., 46: 186-191.

MINCHEW, C.D. & FERGUSON, D.E. (1969) Toxicities of six insecticides to resistant and susceptible green sunfish and golden shiners in static bioassays. J. Mississippi Acad. Sci., 15: 29-32.

MINGELGRIN, U., SALTZMAN, S., & YARON, B. (1977) A possible model for the surface-induced hydrolysis of organophosphorus pesticides on kaolinite clays. Soil Sci. Soc. Am. J., 41: 519-523.

MIRER, F.E., LEVINE, B.S., & MURPHY, S.D. (1977) Parathion and methyl parathion toxicity and metabolism in piperonyl butoxide and diethyl maleate pretreated mice. Chem.-Biol. Interact., 17: 99-112.

MIYAMOTO, J., SATO, Y., KADOTA, T., FUJINAMI, A., & ENDO, M. (1963) Studies on the mode of action of organophosphorus compounds. Part I. Metabolic fate of P^{32} labelled sumithion and methylparathion in guinea pig and white rat. Agric. Biol. Chem., 27: 381-389.

MOELLER, H.C. & RIDER, J.A. (1961) Studies on the anticholinesterase effect of parathion and methylparathion in humans. Fed. Proc., 20: 434.

MOELLER, H.C. & RIDER, J.A. (1962) Threshold of incipient toxicity to systox and methyl parathion. Fed. Proc., 21: 451.

MOHANTY-HEJMADI, P. & DUTTA, S.K. (1981) Effects of some pesticides on the development of the Indian bull frog Rana tigerina. Environ. Pollut., Ser. A, 24: 145-161.

MOHN, G. (1973) 5-methyltryptophan resistance mutations in Escherichia coli K-12. Mutagenic activity of monofunctional alkylating agents including organophosphorus insecticides. Mutat. Res., 20: 7-15.

MOLLHOFF, E. (1981) [Laboratory studies to metabolism of methyl parathion in soils.] Leverkusen, Gemany, Bayer AG (Unpublished report submitted to WHO by Bayer AG) (in German).

MOLNAR, J. & PAKSY, K.A. (1978) [Evaluation of the acute toxicity of inhaled pesticides in experimental animals.] In: [Proceedings of the Conference on Safety Technology in the Agricultural Use of Chemicals.] pp. 1790-1793 (Conference paper) (in German).

MOORTHY, K.S., REDDY, B.K., CHETTY, C.S., & SWAMI, K.S. (1983) Catalytic potential of catalase in hepatopancreas of freshwater mussel Lamellidens marginalis during induced methyl-parathion toxic stress. Geobios, 10(1): 42-44.

MOORTHY, K.S., REDDY, B.K., SWAMI, K.S., & CHETTY, C.S. (1984) Changes in respiration and ionic content in tissues of freshwater mussel exposed to methyl parathion toxicity. Toxicol. Lett., 21: 287-291.

MORGAN, D.P., HETZLER, H.L., SLACH, E.F., & LIN, L.I. (1977) Urinary excretion of paranitrophenol and alkyl phosphates following ingestion of methyl or ethyl parathion by human subjects. Arch. environ. Contam. Toxicol., 6: 159-173.

MORPURGO, G., AULICINO, F., BIGNAMI, M., CONTI, L., VELCICH, A., & MONTALENTI, S.G. (1977) Genetica - Relationship between structure and mutagenicity of dichlorvos and other pesticides. Rend. Cl. Sci. Fis. Mat. Nat., 62: 692-701.

MUDGALL, C.F. & PATIL, H.S. (1987) Toxic effects of dimethoate and methyl parathion on glycogen reserves of male and female Rana. J. Environ. Biol., 8(3): 237-244.

MUHLMANN, R. & SCHRADER, G. (1957) [Hydrolysis of the insecticdal phosphoric acid esters.]. Z. Naturforsch., B12: 196-208 (in German).

MULLER, B. (1973) [Determination of pesticide residues in bee honey. I. Semiquantitative thin-layer chromatographic determination of insecticide residues in bee honey]. Nahrung, 17: 381-386 (in German).

MUNCY, R.J. & OLIVER, A.D. Jr (1963) Toxicity of ten insecticides to the red crawfish *Procambarus clarki*. Trans. Am. Fish. Soc., 96(4): 428-431.

MUNDY, R.L., BOWMAN, M.C., FARMER, J.H., & HALEY, T.J. (1978) Quantitative structure activity study of a series of substituted O,O-dimethyl O-(p-nitrophenyl) phosphorothioates and O-analogs. Arch. Toxicol., 41: 111-123.

MUNN, S., KEEFE, T.J., & SAVAGE, E.P. (1985) A comparative study of pesticide exposures in adult and youth migrant field workers. Arch. environ. Health, 40: 215-220.

MURPHY, S.D. (1980) Toxic interactions with dermal exposure to organophosphate insecticides. Toxicol. Lett., 5: 34.

MURTHY, A. & RAMANI, A. (1982) What is an environmentally safe pesticide? In: Christiansen, K., Koch, B., & Bro-Rasmussen, F., ed. Chemicals in the environment. Chemicals testing and hazard ranking - the interaction between science and administration. Proceedings of the International Symposium, Lyngby-Copenhagen, 18-20 October 1982. Lyngby, The Technical University of Denmark, Laboratory of Environmental Science and Ecology, pp.323-333.

MURTY, A.S., RAMANI, A.V., CHRISTOPHER, K., & RAJABHUSHANAM, B.R. (1984) Toxicity of methyl parathion and fensulfothion to the fish *Mystus cavasius*. Environ. Pollut., A34: 37-46.

NAG, M. & NANDI, N. (1987) *In vitro* and *in vivo* effect of organophosphate pesticides on monoamine oxidase activity in rat brain. Biosci. Rep., 7(10): 801-804.

NAGARATNAMMA, R. & RAMAMURTHI, R. (1982) Metabolic depression in the fresh water teleost *Cyprinus carpio* exposed to an organophosphate pesticide. Curr. Sci., 51(13): 668-669.

NAGYMAJTENYI, L., DESI, I., & LORENCZ, R. (1988) Neurophysiological markers as early signs of organophosphate neurotoxicity. Neurotoxicol. Teratol., 10: 429-434.

NAKATSUGAWA, T., TOLMAN, N.M., & DAHM, P.A. (1968) Degradation and activation of parathion analogs by microsomal enzymes. Biochem. Pharmacol., **17**: 1517-1528.

NAKATSUGAWA, T., TOLMAN, N.M., & DAHM, P.A. (1969) Degradation of parathion in the rat. Biochem. Pharmacol., **18**: 1103-1114.

NANGNIOT, P. (1966) Electrochemical methods for quantitative analysis of pesticide residues. Meded. Fac. Landbouwwet. Rijksuniv., Gent, **31**: 447-473.

NAQVI, S.M.Z. (1973) Toxicity of twenty-three insecticides to a tubificed worm Branchiura sowerbyi from the Mississippi delta. J. Econ. Entomol., **66**(1): 70-74.

NATALE, O.E., GOMEZ, C.E., PECHEN de D'ANGELO, A.M., & SORIA, C.A. (1988) Waterborne pesticides in the Negro River Basin (Argentina). In: Abbou, R., ed. Hazardous waste: detection, control, treatment. Amsterdam, Oxford, New York, Elsevier Science Publishers, pp. 879-907.

NATIONAL FIRE PROTECTION ASSOCIATION (1986) Fire protection guide on hazardous materials, 9th ed. Quincy, Massachusetts, National Fire Protection Association, pp. 49-64.

NATIONAL RESEARCH COUNCIL (1977) Drinking water and health. Washington, DC, National Academy of Sciences, pp. 626-635 (Publication NRC/2619).

NCI (1979) Bioassay of methyl parathion for possible carcinogenicity. Washington, DC, US Department of Health, Education & Welfare, National Cancer Institute (Technical Report Series No. 157; DHEW Publ. No. (NHI) 79-1713).

NEHEZ, M., BOROS, P., FERKE, A., MOHOS, J., PALOTAS, M., VETRO, G., ZIMANYI, M., & DESI, I. (1988) Cytogenetic examination of people working with agrochemicals in the southern region of Hungary. Regul. Toxicol. Pharmacol., **8**: 37-44.

NELSON, R.C. (1967) Procedure for nine organothiophosphate pesticide residues on fruits and vegetables, using microcoulometric gas chromatography. J. Assoc. Off. Anal. Chem., **50**: 922-926.

NEWTON, T.D., GATTIE, D.K., & LEWIS, D.L. (1990) Initial test of the benchmark chemical approach for predicting microbial transformation rates in aquatic environments. Appl. environ. Microbiol., **56**(1): 288-291.

NIELSEN, P.G. (1985) Quantitative analysis of parathion and parathion-methyl by combined capillary column gas chromatography negative ion chemical ionization mass spectrometry. Biomed. Mass Spectrom., **12**: 695-698.

NIETHAMMER, K.R. & BASKETT, T.S. (1983) Cholinesterase inhibition of birds inhabiting wheat fields treated with methyl parathion and toxaphene. Arch. environ. Contam. Toxicol., **12**: 471-475.

NIMMO, W.R., HAMAKER, T.L., MATTHEWS, E., & MOORE, J.C. (1981) An overview of the acute and chronic effects of first and second generation pesticides on an estuarine mysid. In: Vernberg, F.J., Calabrese, A., Thurberg, F.P., & Vernberg, W.B., ed. Biological monitoring of marine pollutants. Proceedings of a Symposium on Pollution and Physiology of Marine Organisms, Milford, Connecticut, 7-9 November 1980. New York, London, Academic Press, pp. 3-19.

NIOSH (1976) Criteria for a recommended standard - occupational exposure to methyl parathion. Cincinnati, Ohio, National Institute for Occupational Safety and Health (DHEW 0 (NIOSH) Publication No. 77-106).

NIOSH (1980) Summary of NIOSH recommendations for occupational health standards. Cincinnati, Ohio, National Institute for Occupational Safety and Health.

NIOSH (1991) Registry of toxic effects of chemical substances. Cincinnatti, Ohio, National Institute for Occupational Safety and Health.

NISHIUCHI, Y. & HASHIMOTO, Y. (1967) Toxicity of pesticide ingredients to some freshwater organisms. Botyu-Kagaku, 32(5): 5-11.

NOMIYAMA, K., MATSUI, K., & NOMIYAMA, H. (1980) Environmental temperature, a factor modifying the acute toxicity of organic solvents, heavy metals, and agricultural chemicals. Toxicol. Lett., 6: 67-70.

OMURA, M., HASHIMOTO, K., OHTA, K., IIO, T., UEDA, S., ANDO, K., & HIRAIDE, H. (1990) Relative retention time diagram as a useful tool for gas chromatographic analysis and electron-capture detection of pesticides. J. Assoc. Off. Anal. Chem., 73(2): 300-306.

OOMEN, P.A. (1986) A sequential scheme for evaluating the hazard of pesticides to bees, *Apis melifera*. Mede. Fac. Landbouwwet. Rijksuniv. Gent, 51(36): 1205-1213.

ORLANDO, E., RAFFI, G.B., CASCELLA, D., DE ROSA, V., & CIPOLLA, C. (1972) Effect of quinidine on electrocardiographic alterations during acute poisoning by phosphoric acid esters. Lav. Um., 24: 1-13.

OSADCHUK, M., ROMACH, M., & MCCULLY, K.A. (1971) Cleanup and separation procedures for multipesticide residue analysis in monitoring and regulatory laboratories. In: Tahori, A.S., ed. Pesticide chemistry. Proceedings of the 2nd International Congress on Pesticide Chemistry, New York, Gordon & Breach, vol. 4., pp. 357-383.

OU, L.T. (1985) Methyl parathion degradation and metabolism in soil: influence of high soil-water contents. Soil Biol. Biochem., 17: 241-243.

OU, L.-T. & SHARMA, A. (1989) Degradation of methyl parathion by a mixed bacterial culture and a *Bacillus* sp. isolated from different soils. J. Agric. Food Chem., 37: 1514-1518.

OU, L.-T., RAO, P.S.C., & DAVIDSON, J.M. (1983) Methyl parathion degradation in soil: influence of soil-water tension. Soil Biol. Biochem., 15: 211-215.

PALAWSKI, D., BUCKLER, D.R., & MAYER, F.L. (1983) Survival and condition of rainbow trout (*Salmo gairdneri*) after acute exposures to methyl parathion, triphenyl phosphate, and DEF. Bull. environ. Contam. Toxicol., 30: 614-620.

PAP, A., SZABO, I., & SZARVAS, F. (1976) Effect of liver damage and of phenobarbital, norandrostenolone, phenyl propionate or chloramphenicol treatment of the poisoning by organic phosphoric acid esters. Kiserl. Orvostud., 28: 172-179.

PASCHAL, D.C., BICKNELL, R., & DRESBACH, D. (1977) Determination of ethyl and methyl parathion in runoff water with high performance liquid chromatography. Anal. Chem., 49: 1551-1554.

PATTERSON, P.L. (1982) New uses of thermionic ionization detectors in gas chromatography. Chromatographia, **16**: 107-111.

PAULUHN, J. (1983) [Methyl parathion (E 120). Studies on the irritant/corrosive effect on skin and eyes (rabbit).] Wuppertal-Elberfed, Bayer AG, Institute of Toxicology (Unpublished report No. 11712, submitted to WHO by Bayer AG, Leverkusen, Germany) (in German).

PETERKA, V. & CERNA, V. (1988) Micellar catalysis in decomposition of some organophosphates. Chem. Prum., **38**(1): 36-39.

PFAENDER, F.K., SHUMAN, M.S., DEMPSEY, H., & HARDEN, C.W. (1977) Monitoring heavy metals and pesticides in the Cape Fear River Basin of North Carolina, Chapel Hill, NC. Raleigh, North Carolina, North Carolina Water Resources Research Institute, pp. 29, 119.

PFEIFFER, R. & STAHR, H.M. (1982) Screening for organochlorine and organophosphorus pesticides. In: Touchstone, J.C., ed. Advances in thin layer chromatography: clinical and environmental applications. Proceedings of the Second Biennial Symposium, Philadelphia, December 1980. New York, John Wiley and Sons, pp. 451-460.

PFLUGMACHER, J. & EBING, W. (1974) Purification of phosphoric acid insecticide residues in vegetable extracts by gel chromatography on Sephadex LH-20. J. Chromatogr., **93**: 457-463.

PICKERING, Q.H., HENDERSON, C., & LEMKE, A.E. (1962) The toxicity of organic phosphorus insecticides to different species of warm-water fishes. Trans. Am. Fish. Soc., **91**: 175-184.

PIONKE, H.B., KONRAD, J.G., CHESTERS, G., & ARMSTRONG, D.E. (1968) Extraction of organochlorine and organophosphate insecticides from lake waters. Analyst, **93**: 363-367.

PLAPP, F.W. & CASIDA, J.E. (1958) Hydrolysis of the alkyl-phosphate bond in certain dialkyl aryl phosphorothioate insecticides by rats, cockroaches, and alkali. J. Econ. Entomol., **51**(6): 800-803.

PORTIER, R.J. & MEYERS, S.P. (1982) Monitoring biotransformation and biodegradation of xenobiotics in simulated aquatic microenvironmental systems. Dev. Ind. Microbiol., **23**: 459-475.

PORTIER, R.J., CHEN, H.M., & MEYERS, S.P. (1983) Environmental effect and fate of selected phenols in aquatic ecosystems using microcosm approaches. Dev. Ind. Microbiol., **24**: 409-424.

PREUSSMANN, R., SCHNEIDER, H., & EPPLE, F. (1969) [Investigation of effects of alkylating agents. II. The investigation of different classes of alkylating agents by means of a modified colour reaction with 4-(4-nitro-benzyl)pyridine (NBP).] Arzneimittelforschung, **7**: 1059-1073 (in German).

PRINSLOO, S.M. & DE BEER, P.R. (1987) Gas chromatographic relative retention data for pesticides on nine packed columns: II. Organophosphorus and organochlorine pesticides, using elecron-capture detection. J. Assoc. Off. Anal. Chem., **70**: 878-888.

PRITCHARD, P.H., CRIPE, C.R., WALKER, W.W., SPAIN, J.C., & BOURQUIN, A.W. (1987) Biotic and abiotic degradation rates of methyl parathion in freshwater and estuarine water and sediment samples. Chemosphere, 16(7): 1509-1520.

QIAN, C., SANDERS, P.F., & SEIBER, J.N. (1985) Accelerated degradation of organophosphorus pesticides with sodium perborate. Bull. environ. Contam. Toxicol., 35: 682-688.

RADULOVIC, L.L., LAFERLA, J.J., & KULKARNI, A.P. (1986) Human placental glutathione s-transferase-mediated metabolism of methyl parathion. Biochem. Pharmacol., 35: 3473-3480.

RADULOVIC, L.L., KULKARNI, A.P., & DAUTERMAN, W.C. (1987) Biotransformation of methyl parathion by human foetal liver glutathione s-transferases: an *in vitro* study. Xenobiotica, 17: 105-114.

RAMAKRISHNA, N. & RAMACHANDRAN, B.V. (1978) Colorimetric determination of fenitrothion and methyl parathion using a hydroxylaminolytic procedure. J. Indian Chem. Soc., LV: 185-187.

RANI, V.J.S., VENKATESHWARLU, P., & JANAIAH, C. (1989) Changes in carbohydrate metabolism of *Clarias bactrachus* (Linn) when exposed to two organophosphorus insecticides. J. environ. Biol., 10: 197-204.

RAO, S.N. & McKINLEY, W.P. (1969) Metabolism of organophosphorus insecticides by Oliver homogenates from different species. Can. J. Biochem., 47: 1155-1159.

RAO, K.S.P. & RAO, K.V.R. (1983) Regulation of phosphorylases and aldolases in tissues of the teleost *Tilapia mossambica* under methyl-parathion impact. Bull. environ. Contam. Toxicol., 31(4): 474-478.

RAO, K.S.P. & RAO, K.V.R. (1984a) Tissue specific alteration of aminotransferases and total ATPases in the fish (*Tilapia mossambica*) under methyl parathion impact. Toxicol. Lett., 20: 53-57.

RAO, K.S.P. & RAO, K.V.R. (1984b) Impact of methyl parathion toxicity and eserine inhibition of acetylcholinesterase activity in tissues of the teleost (*Tilapia mossambica*) - a correlative study. Toxicol. Lett., 22: 351-356.

RAO, K.S.P. & RAO, K.V.R. (1987) The possible role of glucose-6-phosphate dehydrogenase in the detoxification of methyl parathion. Toxicol. Lett., 39: 211-214.

RAO, K.V.R., RAO, K.R.S.S., & RAO, K.S.P. (1983a) Cardiac responses to malathion and methyl parathion in the mussel, *Lamellidens marginalis* (Lamark). J. environ. Biol., 4: 65-68.

RAO, K.S.P., MADHU, C., RAO, K.R.S.S., & RAO, K.V.R. (1983b) Effect of methyl parathion on body weight, water content and ionic changes in the teleost, *Tilapia mossambica* (Peters). J. Food Sci. Technol., 20: 27-29.

RAO, T.S., DUTT, S., & MANGAIAH, K. (1967) TLm values of some modern pesticides to the fresh-water fish *Puntius puckelli*. Environ. Health, 9(2): 103-109.

RASHID, K.A. & MUMMA, R.O. (1984) Genotoxicity of methyl parathion in short-term bacterial test systems. J. environ. Sci. Health, B19: 565-577.

RASTOGI, A. & KULSHRESTHA, S.K. (1990) Effect of sublethal doses of three pesticides on the ovary of a carp minnow *Rasbora daniconius*. Bull. environ. Contam. Toxicol., **45**: 742-747.

RATTNER, B.A. & FRANSON, J.C. (1984) Methyl parathion and fenvalerate toxicity in American kestrels: acute physiological responses and effects of cold. Can. J. Physiol. Pharmacol., **62**: 787-792.

REDDY, K.S. & GAMBRELL, R.P. (1985) The rate of soil reduction as affected by levels of methyl parathion and 2,4-D. J. environ. Sci. Health, **B20**: 275- 298.

REDDY, M.S. & RAO, K.V.R. (1986) Acute toxicity of insecticides to penaeid prawns. Environ. Ecol., **4**(1): 221-223.

REDDY, M.S. & RAO, K.V.R. (1988) *In vivo* recovery of acetylcholinesterase activity from phosphamidon and methylparathion induced inhibition in the nervous tissue of penaeid prawn (*Metapenaeus monoceros*). Bull. environ. Contam. Toxicol., **40**: 752- 758.

REDDY, M.S. & RAO, K.V.R. (1989) *In vivo* modification of lipid metabolism in response to phosphamidon, methylparathion and lindane exposure in the penaeid prawn, *Metapenaeus monoceros*. Bull. environ. Contam. Toxicol., **43**: 603-610.

REDDY, M.S. & RAO, K.V.R. (1990a) Effects of sublethal concentrations of phosphamidon, methyl parathion, DDT, and lindane on tissue nitrogen metabolism in the penaeid prawn, *Metapenaeus monoceros* (Fabricius). Ecotoxicol. environ. Saf., **19**: 47-54.

REDDY, M.S. & RAO, K.V.R. (1990b) Methylparathion - induced alterations in the acetylcholinesterase and phosphatases in a penaeid prawn, *Metapenaeus monoceros* (Fabricius). Bull. Environ. Contam. Toxicol., **45**: 350-357.

REDDY, M.S., RAO, K.V.R., & MURTHY, B.N. (1988) Changes in nitrogen metabolism of penaeid prawn, *Penaeus indicus*, during sublethal phosphamidon and methylparathion-induced stress. Bull. environ. Contam. Toxicol., **41**: 344-351.

REDDY, P.S., BHAGYLAKSHMI, A., & RAMAMURTHI, R. (1985) Moit-inhibition in the crab *Oziotelphusa senex senex* following exposure to malathion and methyl parathion. Bull. environ. Contam. Toxicol., **35**: 92-97.

REDDY, P.S., BHAGYLAKSHMI, A., & RAMAMURTHI, R. (1986a) Carbohydrate metabolism in tissue of fresh water crab (*Oziotelphusa senex senex*) exposed to methyl parathion. Bull. environ. Contam. Toxicol., **36**: 204-210.

REDDY, P.S., BHAGYLAKSHMI, A., & RAMAMURTHI, R. (1986b) Changes in acid phosphatase activity in tissues of crab, *Oziotelphusa senex senex*, following exposure to methyl parathion. Bull. environ. Contam. Toxicol., **37**: 106-112.

REDDY, T.N. & REDDY, S.J. (1989) Voltammetric behaviour of some nitro group containing pesticides. Indian J. environ. Protect., **9**(8): 592-594.

REHWOLDT, R.E., KELLEY, E., & MAHONEY, M. (1977) Investigations into the acute toxicity and some chronic effects of selected herbicides and pesticides on several fresh water fish species. Bull. environ. Contam. Toxicol., **18**(3): 361-365.

RENHOF, M. (1984) [Methyl parathion (active principle of Folidol M): Tests for embryotoxic effects in rabbits after oral administration.] Wuppertal-Elberfeld, Bayer AG, Institute of toxicology (Unpublished reports No. 12907 and 16331, submitted to WHO by Bayer AG, Leverkusen, Germany) (in German).

RENVALL, S., LINSKOG, E., & CLEMENTZ, E. (1975) Residues of organophosphorus pesticides in fruits and vegetables on the Swedish market from July 1967 to April 1973. Presented at the IUPAC Third International Congress of Pesticide Chemistry, Helsinki, 3-9 July 1974. Environ. Qual. Saf., **Suppl.**, **3**: 166-174.

RICE, C.P., OLNEY, C.E., & BIDLEMAN, T.F. (1977) Use of polyurethane foam to collect trace amounts of chlorinated hydrocarbons and other organics from air. WMO In: Air pollution measurement techniques. Geneva, World Meteorological Organization, Part II, pp. 216-224 (Special Environmental Report No. 10; Publication WMO-No. 460).

RICHTER, E.D., ROSENVALD, Z., KASPI, L., LEVY, S., & GRUENER, N. (1986) Sequential cholinesterase tests and symptoms for monitoring organophosphate absorption in field workers and in persons exposed to pesticide spray drift. Toxicol. Lett., **33**: 25-35.

RIDER, J.A., MOELLER, H.C., PULETTI, E.J., & SWADER, J.I. (1969) Toxicity of parathion, systox, octamethyl pyrophosphoramide, and methyl parathion in man. Toxicol. appl. Pharmacol., **14**: 603-611.

RIDER, J.A., SWADER, J.I., & PULETTI, E.J. (1970) Methyl parathion and guthion anticholinesterase effects in human subjects. Fed. Proc., **29**: 349 (Abstr. No.588).

RIDER, J.A., SWADER, J.I., & PULETTI, E.J. (1971) Anticholinesterase toxicity studies with methyl parathion, guthion and phosdrin in human subjects. Fed. Proc., **30**: 1382.

RIPLEY, B.D. & BRAUN, H.E. (1983) Retention time data for organochlorine, organophosphorus, and organonitrogen pesticides on SE-30 capillary column and application of capillary gas chromatography to pesticide residue analysis. J. Assoc. Off. Anal. Chem., **66**: 1084-1095.

RIPPEL, A., KOVAC, J., & SZOKOLAY, A. (1970) [Investigation of the persistence of organophosphorus insecticides in citrus juices.] Nahrung, **14**(3): 223-228 (in German).

ROACH, J.A.G. & ANDRZEJEWSKI, D. (1987) Analysis for pesticide residues by collision-induced fragmentation. In: Rosen, J.D., ed. Application of new mass spectrometry techniques in pesticide chemistry. New York, Chichester, Brisbane, Toronto, John Wiley and Sons, pp. 187-210 (Chemical Analysis series, Volume 91).

ROARK, B., PFRIMMER, T.R., & MERKL, M.E. (1963) Effects of some formulations of methyl parathion, toxaphene and DDT on the cotton plant. Crop Sci., **3**: 338-341.

ROBERTS, D.K., SILVEY, N.J., & BAILEY, E.M. Jr (1988) Brain acetylcholinesterase activity recovery following acute methyl parathion intoxication in two feral rodent species: comparison to laboratory rodents. Bull. environ. Contam. Toxicol., **41**: 26-35.

ROBINSON, S.C., ENDALL, R.J., & ROBINSON, R. (1988) Effects of agricultural spraying of methyl parathion on cholinesterase activity and reproductive success in wild starlings (*Sturnus vulgaris*). Environ. Toxicol. Chem., **7**(5): 343-349.

RODNITZKY, R.L., LEVIN, H.S., & MORGAN, D.P. (1978) Effects of ingested parathion on neurobehavioral functions. Clin. Toxicol., **13**: 347-359.

ROSS, B. & HARVEY, J. (1981) A rapid, inexpensive, quantitative procedure for the extraction and analyses of Penncap-M (methyl parathion) from honeybees (*Apis mellifera* L.), beeswax, and pollen. J. Agric. Food Chem., **29**: 1095-1096.

ROYAL SOCIETY OF CHEMISTRY (1986) European directory of agrochemical products. Volume 3: Insecticides and acaricides. Nottingham, The Royal Society of Chemistry.

RUPA, D.S., REDDY, P.P., & REDDI, O.S. (1989) Chromosomal aberrations in peripheral lymphocytes of cotton field workers exposed to pesticides. Environ. Res., **49**: 1-6.

RUZICKA, J.H., THOMSON, J., & WHEALS, B.B. (1967) The gas chromatographic determination of organophosphorus pesticides. J. Chromatogr., **31**: 37-47.

SABHARWAL, A.K. & BELSARE, D.K. (1986) Persistence of methyl parathion in a carp rearing pond. Bull. environ. Contam. Toxicol., **37**: 705-709.

SADAR, M.H., KUAN, S.S., & GUIBAULT, G.G. (1970) Trace analysis of pesticides using cholinesterase from human serum, rat liver, electric eel, bean leaf beetle, and white fringe beetle. Anal. Chem., **42**: 1770-1774.

SAGREDOS, A.N. & ECKERT, W.R. (1976) [Methods for the determination of phytopharmaceuticals in tobacco and tobacco products. Part II. Simultaneous determination of hexane-soluble organohosphorus pesticides.] Beitr. Tabakforsch., **8**: 447-454 (in German).

SALTZMAN, S., MINGELGRIN, U., & YARON, B. (1976) Role of water in the hydrolysis of parathion and methylparathion on kaolinite. J. Agric. Food Chem., **24**(4): 739-743.

SANDERS, P.F. & SEIBER, J.N. (1983) A chamber for measuring volatilization of pesticides from model soil and water disposal systems. Chemosphere, **12**: 999-1012.

SANDERSON, D.M. & EDSON, E.F. (1964) Toxicological properties of the organophosphorus insecticide dimethoate. Br. J. ind. Med., **21**: 52-64.

SAROJA-SUBBARAJ, G. & BOSE, S. (1982) Effects of methyl parathion on the growth, cell size, pigment and protein content of *Chlorella protothecoides*. Environ. Pollut., **27**: 297.

SAROJA-SUBBARAJ, G. & BOSE, S. (1983a) Correlation between inhibition of photosynthesis and growth of *Chlorella* treated with methyl parathion. J. Biosci., **5**: 71-78.

SAROJA-SUBBARAJ, G. & BOSE, S. (1983b) Binding characteristics of methyl parathion to photosynthetic membranes of *Chlorella*. Pestic. Biochem. Physiol., **20**: 188-193.

SAROJA-SUBBARAJ, G. & BOSE, S. (1984) Recovery from methyl parathion-induced damage of the photosynthetic apparatus in *Chlorella protothecoides*. Bull. Environ. Contam. Toxicol., **32**: 102-108.

SASTRY, C.S.P. & VIJAYA, D. (1986) Spectrophotometric determination of fenitrothion and methyl parathion in insecticidal formulations. J. Food Sci. Technol., **23**: 336-338.

SASTRY, C.S.P. & VIJAYA, D. (1987) Spectrophotometric determination of some insecticides with 3-methyl-2-benzothiazolinone hydrazone hydrochloride. Talanta, **34**: 372-374.

SAUVEGRAIN, P. (1980) Les micropolluants organiques dans les eaux superficielles continentales. Rapport No. 1: Les pesticides organophosphorés. Paris, Bureau National de l'Information Scientifique et Technique.

SAXTON, W.L. (1987) Emergence temperature indexes and relative retention times of pesticides and industrial chemicals determined by linear programmed temperature gas chromatography. J. Chromatogr., **393**: 175-194.

SAYRE, I.M. (1988) International standards for drinking water. J. Am. Water Works Assoc., **80**: 53-60.

SCHAFER, E.W. Jr (1972) The acute oral toxicity of 369 pesticidal, pharmaceutical and other chemicals to wild birds. Toxicol. appl. Pharmacol., **21**: 315-330.

SCHILDE, B. & BOMHARD, E. (1984) [E 605-methyl (methyl parathion). Supplementary histopathological test, further to the 2-year feeding trial on rats. Addendum to report No. 9889.] Wuppertal-Elberfeld, Bayer AG, Institute of Toxicology (Unpublished report No. 12559, submitted to WHO by Bayer AG, Leverkusen, Germany) (in German).

SCHIMMEL, S.C., GARNAS, R.L., PATRICK, J.M. Jr, & MOORE, J.C. (1983) Acute toxicity, bioconcentration and persistence of AC 222.705, benthiocarb, chlorpyrifos, fenvalerate, methylparathion and permethrin in the estuarine environment. J. Agric. Food Chem., **31**(1): 104-113.

SCHNORBUS, R.R. & PHILLIPS, W.F. (1967) New extraction system for residue analyses. J. Agric. Food Chem., **15**: 661-666.

SCHOMBURG, C.J., GLOTFETTY, D.E., & SEIBER, J.N. (1991) Pesticide occurrence and distribution in fog collected near Monterey, California. Environ. Sci. Technol., **25**: 155-160.

SCHOOR, W.P. & BRAUSCH, J. (1980) The inhibition of acetylcholinesterase activity in pink shrimp *Penaeus duorarum* by methyl-parathion and its oxon. Arch. environ. Contam. Toxicol., **9**: 599-605.

SCHRADER, G. (1963) [The development of new insecticidal phosphoric acid esters.] Weinheim, Verlag Chemie GmbH, pp. 223-258 (in German).

SCHULTEN, H.R. & SUN, S. E. (1981) High-resolution field desorption mass spectrometry. Part IX. Field desorption mass spectrometry of standard organophosphorus pesticides and their identification in waste water. Int. J. Environ. Anal. Chem., **10**: 247-263.

SCHULZ, H., DESI, I., & NAGYMAJTENYI, L. (1990) Behavioural effects of subchronic intoxication with parathion-methyl in male Wistar rats. Neurotoxicol. Teratol., **12**: 125-127.

SCHULZE, J.A., MANIGOLD, D.B., & ANDREWS, F.L. (1973) Pesticides in selected Western streams - 1968-71. Pestic. Monit. J., **7**: 73-84.

SCHUTZ, H. & SCHINDLER, A. (1974) [Application of reaction chromatography to chemical-toxicological analysis. Microchemical detection and separation of sixteen nitropesticides by thin-layer chromatography.] Fresenius Z. Anal. Chem., **270**: 356-359 (in German).

SCHUTZMANN, R.L., WOODHAM, D.W., & COLLIER, C.W. (1971) Removal of sulfur in environmental samples prior to gas chromatographic analysis for pesticide residues. J. Assoc. Off. Anal. Chem., **54**: 1117-1119.

SEIBER, J.N., McCHESNEY, M.M., & WOODROW, J.E. (1989) Airborne residues resulting from use of methyl parathion, molinate and thiobencarb on rice in the Sacramento Valley, California. Environ. Toxicol. Chem., **8**: 577-588.

SHAFIK, M.T. & ENOS, H.F. (1969) Determination of metabolic and hydrolytic products of organophosphorus pesticide chemicals in human blood and urine. J. Agric. Food Chem., **17**: 1186-1189.

SHARMA, R.P. & REDDY, R.V. (1987) Toxic effects of chemicals on the immune system. In: Haley, T.J. & Berndt, W.O., ed. Handbook of toxicology. Washington, DC, Hemisphere Publishing Corporation, pp. 555-591.

SHARMA, V.K. JADHAV, R.K., RAO, G.J., SARAF, A.K., & CHANDRA, H. (1990) High performance liquid chromatographic method for the analysis of organophosphorus and carbamate pesticides. Forens. Sci. Int., **48**: 21-25.

SHARMILA, M., RAMANAND, K., ADHYA, T.K., & SETHUNATHAN, N. (1988) Temperature and the persistence of methyl parathion in a flooded soil. Soil Biol. Biochem., **20**: 399-401.

SHARMILA, M., RAMANAND, K., & SETHUNATHAN, N. (1989a) Hydrolysis of methyl parathion in a flooded soil. Bull. environ. Contam. Toxicol., **43**: 45-51.

SHARMILA, M., RAMANAND, K., & SETHUNATHAN, N. (1989b) Effect of yeast extract on the degradation of organophosphorus insecticides by soil enrichment and bacterial cultures. Can. J. Microbiol., **35**: 1105-1110.

SHERMA, J. & BRETSCHNEIDER, W. (1990) Determination of organophosphorus insecticides in water by C-18 solid phase extraction and quantitative TLC. J. liq. Chromatogr., **13**(10): 1983-1989.

SHERMA, J. & SHAFIK, T.M. (1975) Multiclass, multiresidue analytical method for determining pesticide residues in air. Arch. environ. Contam. Toxicol., **3**: 55-71.

SHIHIDO, T. & FUKAMI, J. (1963) Studies on the selective toxicities of organic phosphorous insecticides (II), the degradation of ethyl parathion, methyl parathion, methyl paraoxon, and sumithion in mammal, insect and plant. Botyu-Kagaku, **28**: 69-76.

SHIRES, S.W. (1985) A comparison of the effects of cypermethrin, parathio-methyl and DDT on cereal aphids, predatory beetles, earthworms and litter decomposition in spring wheat. Crop Prot., **4**: 177-193.

SIMMON, V.F., MITCHELL, A.D., & JORGENSON, T.A. (1977) Evaluation of selected pesticides as chemical mutagens *in vitro* and *in vivo* studies. Menlo Park, California, Stanford Research Institute, 251 pp. (PB-268647).

SINGH, K.S. & SINGH, A.P. (1978) Persistence of methyl and ethyl parathion in soil and plant. Pesticides, **12**: 39-41.

SINGH, N.N. & SRIVASTAVA, A.K. (1982) Toxicity of a mixture of aldrin and formothion and other organophosphorus, organochlorine and carbamate pesticides to the Indian catfish *Heteropneustes fossilis*. Comp. Physiol. Ecol., **7**(2): 115-118.

SINGH, S., LEHMANN-GRUBE, B., & GOEDDE, H.W. (1984) Cytogenetic effects of paraoxon and methyl-parathion on cultured human lymphocytes: SCE, clastogenic activity and cell cycle delay. Int. Arch. occup. environ. Health, **54** (3): 195-200.

SMART, E. & STEVENSON, J.H. (1982) Laboratory estimation of toxicity of pyrethroid insecticides to honey bees: relevance to hazard in the field. Bee World, **63**: 150-152.

SMITH, D.C., LEDUC, R., & TREMBLAY, L. (1975) Pesticide residues in the total diet in Canada. IV. 1972 and 1973. Pestic. Sci., **6**: 75-82.

SMITH, J.H., MABEY, W.R., BOHONOS, N., HOLT, B.R., LEE, S.S., CHOU, T.-W., BOMBERGER, D.C., & MILL, T. (1978) Environmental pathways of selected chemicals in freshwater systems, Part II: Laboratory studies. Athens, Georgia, US Environmental Protection Agency (PB288511).

SMITH, S., WILLIS, G.H., MCDOWELL, L.L., & SOUTHWICK, L.M. (1987) Dissipation of methyl parathion and ethyl parathion from cotton foliage as affected by formulation. Bull. environ. Contam. Toxicol., **39**: 280-285.

SMYTH, M.R. & OSTERYOUNG, J.G. (1978) A pulse polarographic investigation of parathion and some other nitro-containing pesticides. Anal. chim. Acta, **96**: 335-344.

SOBTI, R.C., KRISHAN, A., & PFAFFENBERGER, C.D. (1982) Cytokinetic and cytogenetic effects of some agricultural chemicals on human lymphoid cells *in vitro*: organophosphates. Mutat. Res., **102**: 89-102.

SOLON, J.M. & NAIR, J.H. (1970) Effect of a sublethal concentration of LAS (linear alkyl benzene sulfonate detergent) on the acute toxicity of various phosphate pesticides to the fathhead minnow (*Pimephales promelas*). Bull. environ. Contam. Toxicol., **5**(5): 408-413.

SOMLYAY, I.M., VARNAGY, L.E., & PAVLISCSAK, Cs. (1989) Effect of various pesticides on plasma biochemistry of chicken embryos. Mede. Fac. Landbouwwet. Rijksuniv. Gent, **54**(2a): 181-184.

SONOBE, H., SAKUMA, S., YAMADA, K., KAWASAKI, M., KATAYAMA, H., & KUROIWA, Y. (1982) Analysis systems for trace compounds found in barley, malt and hops. In: Proceedings of the 17th Convention, Perth, 7-12 March 1982. Sydney, Institute of Brewing, Australia and New Zealand Section, pp. 56-69.

SPECHT, W. (1978) [Methods for the rapid work-up of large fat quantities for analysis of pesticide residues.] Lebensmittelchem Gerichtl. Chem., **32**: 51-53 (in German).

SPECHT, W. & TILLKES, M. (1980) [Gas chromatographic determination of pesticide residues after cleanup by gel chromatography and mini silice gel column chromatographs.] Fresenius Z. Anal. Chem., **301**: 300-307 (in German).

SPENCER, E.Y. (1982) Guide to the chemicals used in crop protection, 7th ed. Ottawa, Agriculture Canada Research Institute, 394 pp. (Publication No. 1093).

SPINGARN, N.E., NORTHINGTON, D.J., & PRESSELY, T. (1982) Analysis of nonvolatile organic hazardous substances by GC/MS. J. Chromatogr. Sci., **20**: 571-574.

SRIVASTAVA, A.K. & SINGH, N.N. (1981) Effects of exposure to methyl-parathion on carbohydrate metabolism in Indian catfish *Heteropneustes fossilis*. Acta pharmacol. Toxicol., **48**(1): 26-31.

STAHR, H.M., GAUL, M., HYDE, W., & MOORE, R. (1979) A cellulose column cleanup for organophosphorus pesticides. Microchem J., **24**: 97-101.

STAMPER, C.R., BALDUINI, W., MURPHY, S.D., & COSTA, L.G. (1988) Behavioural and biochemical effects of postnatal parathion exposure in the rat. Neurotoxicol. Teratol., **10**: 261-266.

STAN, H.J. & GOEBEL, H. (1983) Automated capillary gas chromatographic analysis of pesticide residues in food. J. Chromatogr., **268**: 55-69.

STAN, H. & GOEBEL, H. (1984) Evaluation of automated splitless and manual on-column injection techniques using capillary gas chromatography for pesticide residue analysis. J. Chromatogr., **314**: 413-420.

STAN, H.J. & MROWETZ, D. (1983) Residue analysis of organophosphorus pesticides in food with two-dimensional gas chromatography using capillary columns and flame photometric detection. J. high Resolut. Chromatogr. Chromatogr. Commun., **6**: 255-263.

STAN, H.J. & MUELLER, H.M. (1988) Evaluation of automated and manual hot-splitless, cold-splitless (PTV), and on-column injection technique using capillary gas chromatography for the analysis for organophosphorus pesticides. J. high Resolut. Chromatogr. Chromatogr. Commun., **11**: 140-143.

STANLEY, C.W., BARNEY, J.E., II, HELTON, M.R., & YOBS, A.R. (1971) Measurement of atmospheric levels of pesticides. Environ. Sci. Technol., **5**: 430-435.

STEINWANDTER, H. (1988) Contributions to the application of gel chromatography in residue analysis. II. A new gel chromatographic system using acetone for the separation of pesticide residues and industrial chemicals. Fresenius Z. Anal. Chem., **331**: 499-502.

STEPHENSON, R.R. & KANE, D.F. (1984) Persistence and effects of chemicals in small enclosures in ponds. Arch. environ. Contam. Toxicol., **13**: 313-326.

STOLL, K. (1982) Residue breakdown in stored fruit. In: Proceedings of the 21th International Horticultural Congress, Hamburg, 29 August-4 September 1982. The Hague, International Society for Horticultural Science, pp. 231-235.

STORHERR, R.W., MURRAY, E.J., KLEIN, I., & ROSENBERG, L.A. (1967) Sweep destillation cleanup of determining of organophosphate and chlorinated hydrocarbon pesticides. J. Assoc. Off. Anal. Chem., **50**: 605-615.

STREET, J.C. & SHARMA, R.P. (1975) Alteration of induced cellular and humoral immune response by pesticides and chemicals of environmental concern: quantitative studies of immunosuppression by DDT, aroclor 1254, carbaryl, carbofuran, and methylparathion. Toxicol. appl. Pharmacol., 32: 587-602.

SULTATOS, L.G. (1987) The role of the liver in mediating the acute toxicity of the pesticide methyl parathion in the mouse. Drug Metab. Dispos., 15(5): 613-617.

SULTATOS, L.G. & WOODS, L. (1988) The role of glutathione in the detoxification of the insecticides methyl parathion and azinphos-methyl in the mouse. Toxicol. appl. Pharmacol., 96: 168-174.

SULTATOS, L.G., KIM, B., & WOODS, L. (1990) Evaluation of estimations *in vitro* of tissue/blood distribution coefficients of organothiophosphate insecticides. Toxicol. appl. Pharmacol., 103: 52-55.

SUPROCK, J.F. & VINOPAL, J.H. (1987) Behavior of 78 pesticides and pesticide metabolites on four different ultra-bond gas chromatographic columns. J. Assoc. Off. Anal. Chem., 70: 1014-1017.

SWAMY, G.S. & VEERESH, A.V. (1987) Methylparathion modulates lipid biosynthesis in sorghum (*Sorghum bicolor* L.). Moench. Pestic. Biochem. Physiol., 28: 341-348.

SWINEFORD, D.M. & BELISLE, A.A. (1989) Analysis of trifluralin, methyl paraoxon, methyl parathion, fenvalerate and 2,4-D dimethylamine in pond water using solid-phase extraction. Environ. Toxicol. Chem., 8: 465-468.

TAKIMOTO, Y., OHSHIMA, M., YAMADA, H., & MIYAMOTO, J. (1984) Fate of fenithrothion in several developmental stages of the killifish (*Oryzias latipes*). Arch. environ. Contam. Toxicol., 13: 579-587.

TANIMURA, T., KATSUYA, T., & NISHIMURA, H. (1967) Embryotoxicity of acute exposure to methyl parathion in rats and mice. Arch. environ. Health, 15: 609-613.

TAYLOR, P. (1980) Anticholinesterase agents. In: Gilman, A.G., Goodman, L.S., & Gilman, A., ed. The pharmacological basis of therapeutics, 6th ed. New York, Macmillan Publishing Co., pp. 100-119.

TESSARI, J.D. & SPENCER, D.L. (1971) Air sampling for pesticides in the human environment. J. Assoc. Off. Anal. Chem., 54: 1376-1382.

THIELEMANN, H. (1974) [Semiquantitative thin layer chromatographic determination of parathion-methyl.] Z. Chem., 14: 407-408 (in German).

THIER, H-P. & ZEUMER, H. (1987) [Pesticides Commission of the German Association for the Encouragement of Research. Manual of pesticide residue analysis.] Weinheim, VCH Verlag, vol. 1.

THOMAS, P.T. & HOUSE, R.V. (1989) Pesticide-induced modulation of the immune system. In: Ragsdale, N.N. & Menzer, R.E., ed. Carcinogenicity and pesticides: principles, issues and relationships. 196th National Meeting of the American Chemical Society, Los Angeles, California, 25-30 September 1988 Washington, DC, American Chemical Society, pp. 94-106 (ACS Symposium series, 414).

THOMPSON, A.R. & GORE, F.L. (1972) Toxicity of twenty-nine insecticides to Folsomia candida: laboratory studies. J. Econ. Entomol., **65**: 1255-1260.

THUMA, N.K., O'NEILL, P.E., BROWNLEE, S.G., & VALENTINE, R.S. (1983) Microbial degradation of selected harzardous materials: pentachlorophenol, hexachlorocyclopentadiene and methyl parathion, Washington, DC, US Environmental Protection Agency, 76 pp. (EPA-Report No. 68-03-2491).

THYSSEN, J. (1979) [E 120 (methyl parathion): Studies on acute inhalation toxicity.] Wuppertal-Elberfeld, Bayer AG, Institute of Toxicology (Unpublished report No. 8148, submitted to WHO by Bayer AG, Leverkusen, Germany) (in German).

THYSSEN, J. & MOHR, U. (1982) [E 120 (methyl parathion): Subacute inhalation test on rats - histopathological findings.] Wuppertal-Elberfeld, Bayer AG, Institute of Toxicology (Unpublished report No. 11302, submitted to WHO by Bayer AG, Leverkusen, Germany) (in German).

TILSTONE, W.J., WINCHESTER, J.F., & REAVEY, P.C. (1979) The use of pharmacokinetics principles in determining the effectiveness of removal of toxins from blood. Clin. Pharmacokinet., **4**: 23-37.

TOSCANO, N.C., SANCES, F.V., JOHNSON, M.W., & LaPRE, L.F. (1982) Effect of various pesticides on lettuce physiology and yield. J. Econ. Entomol., **75**: 738-741.

TRIPATHI, G. & SHUKLA, S.P. (1988) Inhibition of liver and skeletal muscle enzymes by methyl parathion. Biochem. Arch., **4**: 55-61.

TRIPATHI, G. & SHUKLA, S.P. (1990) Enzymatic and ultrastructural studies in a freshwater catfish: impact of methyl parathion. Biomed. Environ. Sci., **3**: 166-182.

TRIPATHY, N.K., DEY, L., & MAJHI, B. (1987) Genotoxicity of metacid established through the somatic and germ line mosai assays and the sex- linked recessive lethal test in *Drosophila*. Arch. Toxicol., **61**(1): 53-57.

US EPA (1981) Acephate, aldicarb, carbophenothion, DEF, EPN, ethoprop, methyl parathion and phorate; their acute and chronic toxicity, bioconcentration potential and persistence as related to marine environments. Gulf Breeze, Florida, US Environmental Protection Agency, Office of Research and Development, Environmental Research Laboratory, 275 pp. (EPA-600/4-81-041; PB 81-244477).

VAN BAO, T, SZABO, I., RUZICSKA, P., & CZEIZEL, R. (1974) Chromosome aberrations in patients suffering acute organic phosphate insecticide intoxication. Humangenetik., **24**: 33-57.

VAN VELD, P.A. & SPAIN, J.C. (1983) Degradation of selected xenobiotic compounds in three types of aquatic test systems. Chemosphere, **12**: 1291- 1305.

VARIS, A.-L. (1972) Loss of lindane, dimethoate, and methyl parathion residues from seedlings of sugar beet as influenced by plant growth. Ann. Agric. Fenn., **11**: 381-385.

VARNAGY, L. & DELI, E. (1985) Comparative teratological study of insecticide Wofatox 50 EC (50% methyl parathion) on chicken and pheasant fetuses. Anat. Anz. Jena, **158**: 1-3.

VARNAGY, L., KORZENSZKY, M., & FANCSI, T. (1984) Teratological examination of the insecticide methylparathion/Wofatox 50 EC/ on pheasant embryos. 1. Morphological study. Vet. Res. Commun., **8**: 131-139.

VARNAGY, L., DELI, E., & BAUMANN, M. (1985) Scanning electron microscopic examination of cartilage in chicken embryos treated with the insecticide Wofatox 50 EC. Acta vet. Hung., **36**:(1-2): 117-121.

VOGELGESANG, J. & THIER, H.P. (1986) [Contributions to the analysis of pesticide residues in ready-to-eat foods.] Z. Lebensm.-Unters. Forsch., **182**(5): 400-406 (in German).

VROCHINSKY, K.K. & MAKOVSKY, V.N. (1977) The sources of pollution and the content of pesticides in air. In: Uses of pesticides and environmental protection. Kiev, Vysshaya Shkola, pp. 143-154.

WALKER, T.W., MEEK, C.L., WRIGHT, V.L., & BILLODEAUX, J.S. (1985) Susceptibility of *Romanomermis culicivorax* (Nematoda: Mermithidae) post parasites to agrochemicals used in Louisiana rice production. J. Am. Control Assoc., **1**: 477-481.

WALKER, W.W. (1978) Insecticide persistence in natural sea water as affected by salinity, temperature and sterility. Washington, DC, US Environmental Protection Agency, Office of Research Development, 25 pp. (EPA-600/3-78-044).

WALSH, G.E. & ALEXANDER, S.V. (1980) A marine algal bioassay method: results with pesticides and industrial wastes. Water, Air, Soil Pollut., **13**: 45-55.

WALSH, G.E., DEANS, C.H., & McLAUGHLIN, L.L. (1987) Comparison of the EC_{50}s of algal toxicity tests calculated by four methods. Environ. Toxicol. Chem., **6**: 767-770.

WALTERS, S.M. (1990) Clean-up techniques for pesticides in fatty foods. Anal. chim. acta, **236**: 77-82.

WARE, G.W., ESTESEN, B., & CAHILL, W.P. (1974a) Dislodgable leaf residues of insecticides on cotton. Bull. environ. Contam. Toxicol., **11**: 434-437.

WARE, G.W., MORGAN, D.P., ESTESEN, B.J., & CAHILL, W.P. (1974b) Establishment of reentry intervals for organophosphate-treated cotton fields based on human data. II. Azodrin, ethyl and methyl parathion. Arch. environ. Contam. Toxicol., **2**(2): 117-129.

WARE, G.W., WATSON, T.F., ESTESEN, B., & BUCK, N.A. (1980) Effects of molasses or toxaphene on residual life and efficacy of methyl parathion on cotton. J. Econ. Entomol., **73**: 15-17.

WARE, G.W., BUCK, N.A., & ESTESEN, B.J. (1983) Dislodgeable insecticide residues on cotton foliage: comparison of ULV/cottonseed oil vs. aqueous dilutions of 12 insecticides. Bull. environ. Contam. Toxicol., **31**: 551-558.

WARNER, J.S. (1975) Identification on impurities in technical-grade pesticides. In: Substitute Chemical Programme - the first year of progress. Proceedings of a symposium. Vol. IV: Chemical methods workshop. Washington DC, Bureau of Commerce, National Technical Information Service, pp. 31-65 (PB 261-007).

WATTS, R.R. & STORHERR, R.W. (1967) Sweep codistillation cleanup of milk for determination of organophosphate and chlorinated hydrocarbon pesticides. J. Assoc. Off. Anal. Chem., **50**: 581-585.

WEBER, H., PATZSCHKE, K., & WEGNER, L.A. (1979) [Methyl parathion-^{14}C (benzene ring-labelled compound): Biokinetic studies on rats.] Wuppertal-Elberfeld, Bayer AG, Pharmokinetics Institute, Isotope Laboratory (Unpublished report No. Pharma-8722, submitted to WHO by Bayer AG, Leverkusen, Germany) (in German).

WEGMAN, R.C.C., VAN DEN BROEK, H.H., HOFSTEE, A.W.M., & MARSMAN, J.A. (1984) Determination of triazines, organophosphorus containing pesticides and aromatic amines in soil samples. Meded. Fac. Landbouwwet. Rijksuniv. Gent., **49**: 1231-1239.

WEHNER, T.A., WOODROW, J.E., KIM, Y.H., & SEIBER, J.N. (1984) Multiresidue analysis of trace organic pesticides in air. In: Keith, L.H., ed. Identification and analysis of organic pollutants. Boston, Butterworth, pp. 273-290.

WERNER, D. & PAWLITZ, H. (1978) Differential elimination of phenol by diatoms and other unicellular algae from low concentrations. Bull. environ. Contam. Toxicol., **20**: 303-312.

WESSEL, J.R. (1967) Collaborative study of a method for multiple organophosporus pesticide residues in nonfatty foods. J. Assoc. Off. Anal. Chem., **50**: 430-439.

WEST, B., VIDONE, L.B., & SHAFFER, C.B. (1961) Acute and subacute toxicity of dimethoate. Toxicol. appl. Pharmacol., **3**: 210-223.

WHITE-STEVENS, R. (1971) Pesticides in the environment, New York, Basel, Marcel Dekker, Inc., vol. 1., parts 1 and 2.

WHO (1990) The WHO recommended classification of pesticides by hazard and guidelines to classification, 1990-91. Geneva, World Health Organization (Unpublished document WHO/PCS/90.1).

WHO (1986) Environmental Health Criteria, 63: Organophosphorus insecticides: a general introduction. Geneva, World Health Organization, 181 pp.

WHO/FAO (1975) Data sheets on pesticides No. 7, Rev. 1: Parathion-methyl. Geneva, World Health Organization (Unpublished WHO document VBC/05/75.7 (Rev. 1)).

WICKER, G.W., WILLIAMS, W.A., & GUTHRIE, F.E. (1979) Exposure of field workers to organophosphorus insecticides: sweet corn and peaches. Arch. environ. Contam. Toxicol., **8**: 175-182.

WIERSMA, G.B., MITCHELL, W.G., & STANFORD, C.L. (1972) Pesticide residues in onions and soil - 1969. Pestic. Monit. J., **5**(4): 345-347.

WILD, D. (1975) Mutagenicity studies on organophosphorus insecticides. Mutat. Res., **32**: 133-150.

WILKINS, J.P.G. (1990) Rationalization of the mass spectrometric and gas chromatographic behaviour of organophosphorus pesticides: Part 1 - Substituted phenyl phosphorothioates. Pest. Sci., **29**: 163-181.

WILLEMS, J., BRAECKMAN, R., & BELPAIRE, F. (1980) Fate of methyl parathion in the dog. Toxicol. Lett., 5(Suppl. 1): 34 (Abstract No. 038).

WILLIAMS, M.W., FUYAD, J.P., FRAWLEY, J.P., & FITZHUGH, O.G. (1957) In vivo effects of paired combinations of five organic phosphate insecticides. J. Agric. Food Chem., 6: 514-516.

WILLIAMS, M.W., FUYAT, H.N., & FITZHUGH, O.G. (1959) The subacute toxicity of four organic phosphates to dogs. Toxicology, 1: 1-7.

WILLIS, G.H., McDOWELL, L.L., SOUTHWICK, L.M., & SMITH, S. (1985) Toxaphene, methyl parathion, and fenvalerate disapperance from cotton foliage in the mid-south. J. environ. Qual., 14(3): 446-450.

WILMES, R. (1987) Parathion-methyl: Hydrolysis studies. Leverkusen, Germany, Bayer AG, Institute of Metabolism Research, 34 pp. (Unpublished report No. PF 2883, submitted to WHO by Bayer AG).

WINDHOLZ, M., BUDVARI, S., BLUMETTI, R.F., & OTTERBEIN, E.S. ed. (1983) The Merck Index, 10th ed. Rahway, New Jersey, Merck and Co., 874 pp.

WINTER, H. & LINDNER, K. (1987) [Drinking water of river Rhine after the "Sandoz accident".] Wasser Abwasser, 128: 525-532 (in German).

WOLFE, H.R., DURHAM, W.F., & ARMSTRONG, J.F. (1967) Exposure of workers to pesticides. Arch. environ. Health, 14: 622-633.

WOLFE, H.R., WASH, W., DURHAM, W.F., & ARMSTRONG, J.F. (1970) Urinary excretion of insecticide metabolites. Excretion of para-nitrophenol and DDA as indicators of exposure to parathion. Arch. environ. Health, 21: 711-716.

WOLFE, N.L., KITCHENS, B.E., MACALDY, D.L., & GRUNDL, T.J. (1986) Physical and chemical factors that influence the anaerobic degradation of methyl parathion in sediment systems. Environ. Toxicol. Chem., 5: 1019-1026.

WOODER, M.F., WRIGHT, A.S., & KING, L.J. (1977) In vivo alkylation studies with dichlorvos at practical use concentrations. Chem.-Biol. Interact., 19: 25-46.

WOODROW, J.E., SEIBER, J.N., CROSBY, D.G., MOILANEN, K.W., SODERQUIST, C.J., & MOURER, C. (1977) Airborne and surface residues of parathion and its conversion products in a treated plum orchard environment. Arch. Environ. Contam. Toxicol., 6: 175-191.

WORTHING, C.R. & HANCE, R.J. (1990) The pesticide manual, 9th ed. Croydon, British Crop Protection Council.

XUE, J. (1984) Resin concentration method for determination of pesticides in drinking water. Huanjing Huaxue, 3: 43-49.

YAMAMOTO, T., EGASHIRA, T., YOSHIDA, T., & KUROIWA, Y. (1983) Comparative metabolism of fenitrothion and methylparathion in male rats. Acta pharmacol. toxicol., 53: 96-102.

YANG, C.-F. & SUN, Y.-P. (1977) Partition distribution of insecticides as a critical factor affecting their rates of absorption from water and relative toxicities to fish. Arch. environ. Contam. Toxicol., **6**: 325-335.

YASOSHIMA, M. & MASUDA, Y. (1986) Effect of carbon disulfide on the anticholinesterase action of several organophosphorus insecticides in mice. Toxicol. Lett., **32**: 179-184.

YASUNO, M., HIRAKOSO, S., SASA, M., & UCHIDA, M. (1965) Inactivation of some organophosphorus insecticides by bacteria in polluted water. Jpn. J. exp. Med., **35**(6): 545-563.

YODER, J., WATSON, M., & BESON, W.W. (1973) Lymphocyte chromosome analysis of agricultural workers during extensive occupational exposure to pesticides. Mutat. Res., **21**: 335-340.

YOUNGMAN, R.R., TOSCANO, N.C., & GASTON, L.K. (1989) Degradation of methyl parathion to p-nitrophenol on cotton and lettuce leaves and its effects on plant growth. J. Econ. Entomol., **82**: 1317-1322.

YOUNGMAN, R.R., LEIGH, T.F., KERBY, T.A., TOSCANO, N.C., & JACKSON, C.E. (1990) Pesticides and cotton: effect on photosynthesis, growth, and fruiting. J. Econ. Entomol., **83**: 1549-1557.

YOUSSEF, S.H.A., EL-SAYED, M.G.A., & ATEF, M. (1987) Influence of gentamicin and rifamycin on toxicity and biotransformation of methyl parathione in rats. Dtsch. Tierärztl. Wochenschr., **94**: 193-236.

ZEPP, R.G. & SCHLOTZHAUER, P.F. (1983) Influence of algae on photolysis rates of chemicals in water. Environ. Sci. Technol., **17**: 462-468.

ZHAO, S. & WANG, Y. (1984) [High-performance liquid chromatographic analysis of trichlorophon and methylparathionmixed powder.] Sepu, **1**: 134-135 (in Chinese).

ZIETEK, M. (1976) [Polarographic determination of parathion and similar insecticides alongside their metabolites in blood without an extraction procedure.] Mikrochim. Acta, **2**: 249-257 (in German).

ZITKO, V. & McLEESE, D.W. (1980) Evaluation of hazards of pesticides used in forest spraying to the aquatic environment. St. Andrews, New Brunswick, Biological Station (Canadian Technical Report, Fish and Aquatic Sciences, No. 985).

ZWEIG, G. & DEVINE, J.M. (1969) Determination of organophosphorus pesticides in water. Residue Rev., **26**: 17-36.

ANNEX I. TREATMENT OF ORGANOPHOSPHATE INSECTICIDE POISONING IN MAN

(From EHC 63: Organophosphorus insecticides - a general introduction)

All cases of organophosphorus poisoning should be dealt with as an emergency and the patient sent to hospital as quickly as possible. Although symptoms may develop rapidly, delay in onset or a steady increase in severity may be seen up to 48 h after ingestion of some formulated organophosphorus insecticides.

Extensive descriptions of treatment of poisoning by organophosphorus insecticides are given in several major references (Kagan, 1977; Taylor, 1980; UK DHSS, 1983; Plestina, 1984) and will also be included in the IPCS Health and Safety Guides to be prepared for selected organophosphorus insecticides.

The treatment is based on:

(a) minimizing the absorption;
(b) general supportive treatment; and
(c) specific pharmacological treatment.

I.1 Minimizing the absorption

When dermal exposure occurs, decontamination procedures include removal of contaminated clothes and washing of the skin with alkaline soap or with a sodium bicarbonate solution. Particular care should be taken in cleaning the skin area where venepuncture is performed. Blood might be contaminated with direct-acting organophosphorus esters and, therefore, inaccurate measures of ChE inhibition might result. Extensive eye irrigation with water or saline should also be performed. In the case of ingestion, vomiting might be induced, if the patient is conscious, by the administration of ipecacuanha syrup (10-30 ml) followed by 200 ml water. This treatment is, however, contraindicated in the case of pesticides dissolved in hydrocarbon solvents. Gastric lavage (with addition of bicarbonate solution or activated charcoal) can also be performed, particularly in unconscious patients, taking care to prevent aspiration of fluids into the lungs (i.e., only after a tracheal tube has been put into place).

The volume of fluid introduced into the stomach should be recorded and samples of gastric lavage frozen and stored for subsequent chemical analysis. If the formulation of the pesticide involved is available, it should also be stored for further analysis (i.e., detection of toxicologically relevant impurities). A purgative can be administered to remove the ingested compound.

I.2 General supportive treatment

Artificial respiration (via a tracheal tube) should be started at the first sign of respiratory failure and maintained for as long as necessary.

Cautious administration of fluids is advised, as well as general supportive and symptomatic pharmacological treatment and absolute rest.

I.3 Specific pharmacological treatment

I.3.1 Atropine

Atropine should be given, beginning with 2 mg iv and given at 15-30-min intervals. The dose and the frequency of atropine treatment varies from case to case, but should maintain the patient fully atropinized (dilated pupils, dry mouth, skin flushing, etc.). Continuous infusion of atropine may be necessary in extreme cases and total daily doses up to several hundred mg may be necessary during the first few days of treatment.

I.3.2 Oxime reactivators

Cholinesterase reactivators (e.g., pralidoxime, obidoxime) specifically restore AChE activity inhibited by organophosphates. This is not the case with enzymes inhibited by carbamates. The treatment should begin as soon as possible, because oximes are not effective on "aged" phosphorylated ChEs. However, if absorption, distribution, and metabolism are thought to be delayed for any reasons, oximes can be administered for several days after intoxication. Effective treatment with oximes reduces the required dose of atropine. Pralidoxime is the most widely available oxime. A dose of 1 g pralidoxime can be given either im or iv and repeated 2-3 times per day or, in extreme cases, more often. If possible, blood samples should be taken for AChE determinations before and during treatment. Skin should be carefully cleansed before sampling. Results of the assays should influence the decision whether to continue oxime therapy after the first 2 days.

There are indications that oxime therapy may possibly have beneficial effects on CNS-derived symptoms.

I.3.3 Diazepam

Diazepam should be included in the therapy of all but the mildest cases. Besides relieving anxiety, it appears to counteract some aspects of CNS-derived symptoms that are not affected by atropine. Doses of 10 mg sc or iv are appropriate and may be repeated as required (Vale & Scott, 1974). Other centrally acting drugs and drugs that may depress respiration are not recommended in the absence of artificial respiration procedures.

I.3.4 Notes on the recommended treatment

I.3.4.1 Effects of atropine and oxime

The combined effect far exceeds the benefit of either drug singly.

I.3.4.2 Response to atropine

The response of the eye pupil may be unreliable in cases of organophosphorus poisoning. A flushed skin and drying of secretions are the best guide to the effectiveness of atropinization. Although repeated dosing may well be necessary, excessive doses at any one time may cause toxic side-effects. Pulse-rate should not exceed 120/min.

I.3.4.3 Persistence of treatment

Some organophosphorus pesticides are very lipophilic and may be taken into, and then released from, fat depots over a period of many days. It is therefore quite incorrect to abandon oxime treatment after 1-2 days on the supposition that all inhibited enzyme will be aged. Ecobichon et al. (1977) noted prompt improvement in both condition and blood-ChEs in response to pralidoxime given on the 11th-15th days after major symptoms of poisoning appeared due to extended exposure to fenitrothion (a dimethyl phosphate with a short half-life for aging of inhibited AChE).

3.4.4 Dosage of atropine and oxime

The recommended doses above pertain to exposures, usually for an occupational setting, but, in the case of very severe exposure or massive ingestion (accidental or deliberate), the therapeutic doses may be extended considerably. Warriner et al. (1977) reported the case of a patient who drank a large quantity of dicrotophos, in error, while drunk. Therapeutic dosages were progressively increased up to 6 mg atropine iv every 15 min together with continuous iv infusion of pralidoxime chloride at 0.5 g/h for 72 h, from days 3 to 6 after intoxication. After considerable improvement, the patient relapsed and further aggressive therapy was given at a declining rate from days 10 to 16 (atropine) and to day 23 (oxime), respectively. In total, 92 g of pralidoxime chloride and 3912 mg of atropine were given and the patient was discharged on the thirty-third day with no apparent sequelae.

References to Annex I.

ECOBICHON, D.J., OZERE, R.L., REID, E., & CROCKER, J.F.S (1977) Acute fenitrothion poisoning. *Can. Med. Assoc. J.*, **116**: 377-379.

KAGAN, JU.S. (1977) [*Toxicology of organophosphorus pesticides,*] Moscow, Meditsina, pp. 111-121, 219-233, 260-269 (in Russian).

PLESTINA, R. (1984) *Prevention, diagnosis, and treatment of insecticide poisoning*, Geneva, World Health Organization (Unpublished document VBC/84.889).

TAYLOR, P. (1980) Anticholinesterase agents. In: Goodman, L.S. & Gilman, A., ed. *The pharmacological basis of therapeutics*, 6th ed., New York, Macmillan Publishing Company, pp. 100-119.

UK DHSS (1983) *Pesticide poisoning: notes for the guidance of medical practitioners*, London, United Kingdom Department of Health and Social Security, pp. 41-47.

VALE, J.A. & SCOTT, G.W. (1974) Organophosphorus poisoning. *Guy's Hosp. Rep.*, **123**: 13-25.

WARRINER, R.A., III, NIES, A.S., & HAYES, W.J., Jr (1977) Severe organophosphate poisoning complicated by alcohol and terpentine ingestion. *Arch. environ. Health*, **32**: 203-205.

RESUME ET EVALUATION, CONCLUSIONS, RECOMMANDATIONS

1 Résumé et évaluation

1.1 Exposition

Le parathion-méthyl est un insecticide organophosphoré dont la première synthèse remonte aux années 1940. Il est relativement insoluble dans l'eau, peu soluble dans l'éther de pétrole et les huiles minérales et facilement soluble dans la plupart des solvants organiques. A l'état pur, il se présente sous la forme de cristaux blancs; le parathion-méthyl technique est légèrement jaunâtre et dégage une odeur alliacée. Il est instable à la chaleur et se décompose rapidement au-dessus de pH 8.

La chromatographie en phase gazeuse avec détection par ionisation de flamme alcaline (AFID) ou photométrie de flamme (FPD) est la méthode la plus couramment utilisée pour le dosage du parathion-méthyl. Les limites de détection dans l'eau vont de 0,01 à 0,1 μg/litre; dans l'air, elles vont de 0,1 à 1 ng/m^3. La chromatographie en phase liquide à haute performance et la chromatographie en couche mince sont également utiles pour la recherche du parathion-méthyl.

La distribution du parathion-méthyl dans l'air, l'eau, le sol et les êtres vivants dépend de plusieurs facteurs physiques, chimiques et biologiques.

Les études utilisant des modèles d'écosystèmes ainsi que des modèles mathématiques montrent que le parathion-méthyl se partage principalement entre l'air et le sol dans l'environnement, une plus faible proportion se répartissant entre les végétaux et les animaux. Il ne se déplace pratiquement pas dans le sol et ni le composé initial, ni ses produits de décomposition n'atteignent normalement les eaux souterraines. Le parathion-méthyl présent dans l'air provient principalement de l'épandage de ce composé, encore qu'il puisse se volatiliser en partie lorsque l'eau qui le contient s'évapore de la surface des feuilles et du sol. Les niveaux atmosphériques de fond dans les zones agricoles vont de zéro (non décelable) à environ 70 ng/m^3. Les concentrations dans l'air après épandage diminuent rapidement en trois jours pour atteindre le niveau de fond au bout d'environ neuf jours. Dans les cours d'eau, les concentrations (études de laboratoire) tombent à 80% de la concentration initiale au bout d'une heure et à 10 % au bout d'une semaine. Le parathion-méthyl demeure plus longtemps dans le sol que dans l'air ou l'eau encore que sa rétention dépende en grande partie du type de sol; dans les sols sableux, les résidus de parathion-méthyl disparaissent plus rapidement que dans le terreau. Les résidus présents à la surface des plantes ou dans les feuilles diminuent rapidement avec une demi-vie de l'ordre de quelques heures; la disparition totale du parathion-méthyl s'effectue en six à sept jours environ.

L'organisme animal est capable de décomposer le parathion-méthyl et d'en éliminer les produits de dégradation en très peu de temps. Ce processus est plus lent chez les vertébrés inférieurs et les invertébrés que chez les mammifères et les oiseaux. Les facteurs de bioconcentration sont faibles et le parathion-méthyl ne s'accumule que temporairement.

C'est la dégradation microbienne qui est de loin la voie la plus importante de dégradation du parathion-méthyl dans le milieu. Le composé disparaît plus rapidement sur le terrain ou dans des modèles d'écosystèmes que ne l'avaient laissé entrevoir les études de laboratoire. Cela tient au fait qu'il existe plusieurs microorganismes capables de décomposer cette substance dans diverses circonstances et dans différents biotopes. La présence de sédiments ou de surfaces végétales qui accroît les populations microbiennes, augmente la vitesse de décomposition du parathion-méthyl.

Sous l'action du rayonnement ultra-violet ou de la lumière solaire, le parathion-méthyl peut subir une décomposition oxydante en paraoxon-méthyl, moins stable; après pulvérisation, le temps de demi-décomposition par le rayonnement ultra-violet est d'environ 40 heures. Toutefois, la contribution de la photolyse à l'élimination totale dans un système aquatique, n'est, selon les estimations, que de 4 %. L'hydrolyse du parathion-méthyl se produit également plus rapidement en milieu alcalin. Une forte salinité favorise aussi l'hydrolyse. En présence de sédiments fortement réducteurs, on a noté des demi-vies de quelques minutes, encore que la sorption à d'autres sédiments accroisse la stabilité du composé.

Dans des villes situées au centre de zones agricoles des Etats-Unis d'Amérique, on a observé que les concentrations de parathion-méthyl dans l'air variaient avec la saison et culminaient en août ou septembre; les enquêtes ont révélé que les teneurs maximales se situaient principalement dans les limites de 100 à 800 ng/m^3 au cours de la période de végétation. Dans les eaux naturelles de ces mêmes régions des Etats-Unis, on a observé des concentrations allant jusqu'à 0,46 μg/litre, les maxima étant atteints en été. Il n'existe qu'un petit nombre de publications sur les résidus alimentaires de parathion-méthyl dans le monde. Aux Etats-Unis, ces résidus se situent en général à un très faible niveau, même si quelques échantillons dépassent les limites maximales de résidus (LMR). Les études de ration totale dont il est fait état dans la littérature ne font état que de traces de résidus. C'est dans les légumes-feuilles (jusqu'à 2 mg/kg) et les légumes racines (jusqu'à 1 mg/kg) que l'on a constaté les résidus les plus élevés lors d'enquêtes sur le panier de la ménagère effectuées aux Etats-Unis entre 1966 et 1969. La préparation, la cuisson et la conservation des aliments entraînent la décomposition des résidus de parathion-méthyl, ce qui réduit encore l'exposition des consommateurs. En cas d'erreurs de manipulation du parathion-méthyl, on peut trouver des résidus plus élevés dans les légumes et les fruits crus. la production, la formulation, la manipulation et l'utilisation du parathion-méthyl comme insecticide sont les principales sources potentielles d'exposition humaine. C'est principalement par contact cutané et, dans une moindre proportion, par inhalation que les travailleurs sont exposés à cette substance.

Lors d'une étude sur des ouvriers agricoles qui pulvérisaient du parathion-méthyl (les ouvriers non protégés procédant à un épandage manuel de cette substance à très bas volume), on a calculé que ces personnes absorbaient 0,4 à 13 mg de parathion-méthyl par 24 heures en se fondant sur le dosage du *p*-nitrophénol dans les urines. Si les ouvriers reviennent trop tôt sur les lieux après le traitement, ils se trouvent encore davantage exposés.

La population générale peut être exposée à des résidus présents dans l'air, l'eau et les aliments par suite de traitements sur les cultures ou les forêts ou d'erreurs de manipulation (épandage en dehors de la zone à traiter) qui entraînent la contamination des champs, des cultures, de l'eau et de l'air.

1.2 Fixation, métabolisme et excrétion

Le parathion-méthyl est facilement absorbé par toutes les voies d'exposition (orale, percutanée, respiratoire) et il se répand rapidement dans les tissus de l'organisme. Les concentrations maximales dans les divers organes ont été observées une à deux heures après le traitement. La conversion du parathion-méthyl en paraoxon-méthyl se produit dans les minutes qui suivent l'administration. Après administration de parathion-méthyl par voie intraveineuse à des chiens, on a observé une demi-vie terminale moyenne de 7,2 heures. C'est le foie qui joue le principal rôle dans le métabolisme et la détoxication du parathion-méthyl. Le mode principal de détoxication du parathion-méthyl et du paraoxon-méthyl au niveau du foie consiste en oxydation, hydrolyse et déméthylation ou désarylation en présence de glutathion réduit (GSH). Les produits de réaction sont le thiophosphate de *o*-méthyle et de *o*-nitrophényle ainsi que les acides diméthylphosphorothioïque ou diméthyl-phosphorique et le *p*-nitrophénol. Il est donc possible d'évaluer l'exposition en mesurant l'excrétion urinaire du *p*-nitrophénol. Chez des volontaires, l'excrétion urinaire de *p*-nitrophénol était de 60 % quatre heures après l'administration et d'environ 100 % au bout de 24 heures. Le métabolisme du parathion-méthyl joue un rôle important dans la toxicité sélective de ce composé pour les différentes espèces et l'apparition éventuelle d'une résistance. L'élimination du parathion-méthyl et de ses métabolites s'effectue principalement par la voie urinaire. Des études menées sur des souris avec du parathion-méthyl radiomarqué au ^{32}P ont montré qu'au bout de 72 heures, 75 % de la radio-activité se retrouvaient dans les urines et jusqu'à 10 % dans les matières fécales.

1.3 Effets sur les êtres vivants dans leur milieu naturel

Certains microorganismes peuvent utiliser le parathion-méthyl comme source de carbone et l'étude d'une communauté naturelle a montré que des concentrations allant jusqu'à 5 mg/litre augmentaient la biomasse et l'activité reproductrice. L'effet est positif dans le cas des bactéries et des actinomycètes; par contre, les champignons et les levures sont moins capables d'utiliser ce composé. Chez une diatomée, on a constaté une inhibition de 50 % de la croissance à une concentration d'environ 5 mg/litre. Chez des algues vertes unicellulaires, la croissance a été réduite par des concentrations comprises entre 25 et 80 μg de parathion-méthyl par litre. Les populations d'algues devenaient tolérantes au parathion-méthyl après quelques semaines d'exposition.

Le parathion-méthyl est extrêmement toxique pour les invertébrés aquatiques, la CL_{50} étant plupart du temps comprise entre < 1 μg et environ 40 μg/litre. Quelques espèces d'arthropodes sont moins sensibles. Pour la daphnie (*Daphnia magna*) la concentration sans effet observable est de 1,2 μg/litre. Les mollusques sont beaucoup moins sensibles, puisque leur CL_{50} varie de 12 à 25 mg/litre.

La plupart des espèces de poissons d'eau douce ou de mer ont une CL_{50} comprise entre 6 et 25 mg/litre, quelques espèces étant nettement plus ou nettement moins sensibles au composé. La toxicité aiguë est comparable chez les amphibiens et les poissons.

Le traitement au parathion-méthyl de mares expérimentales a permis d'en observer les effets sur l'effectif des communautés d'invertébrés aquatiques. Seul un épandage sur les étendues d'eau serait susceptible d'engendrer les concentrations nécessaires à l'apparition de ces effets et encore, seraient-ils de courte durée. Une décimation des populations d'invertébrés est donc improbable en situation réelle. En cas de mortalité chez les invertébrés, les effets ne seraient probablement pas de longue durée.

Il convient dont de veiller à ne pas procéder à des épandages sur les mares, cours d'eau et lacs. Le parathion-méthyl ne doit jamais être épandu lorsque le vent souffle.

Le parathion-méthyl est un insecticide non-sélectif qui détruit les espèces utiles tout autant que les ravageurs. On a fait état de mortalité parmi des abeilles à la suite d'épandages de parathion-méthyl. Ce genre d'accidents est plus grave avec le parathion-méthyl qu'avec d'autres insecticides. Les abeilles adaptées à l'Afrique supportent mieux le parathion-méthyl que les souches européennes.

Le parathion-méthyl s'est révélé modérément toxique pour les oiseaux au laboratoire, la DL_{50} aiguë par voie orale allant de 3 à 8 mg/kg de poids corporel. Par la voie alimentaire, la CL_{50} allait de 70 à 680 mg/kg de nourriture. Rien n'indique que les oiseaux aient à souffrir du parathion-méthyl lorsqu'il est épandu conformément aux recommandations.

On veillera tout particulièrement à l'horaire des épandages pour éviter tout effet nocif sur les abeilles.

1.4 Effets sur les animaux d'expérience et les systèmes d'épreuve in vitro

La DL_{50} par voie orale varie chez les rongeurs de 3 à 35 mg/kg de poids corporel et la DL_{50} par voie percutanée, de 44 à 67 mg/kg de poids corporel.

L'intoxication par le parathion-méthyl engendre les effets cholinergiques habituels des organophosphorés que l'on peut attribuer à l'accumulation d'acétylcholine au niveau des terminaisons nerveuses. La toxicité du parathion-méthyl est due à sa métabolisation en paraoxon-méthyl. Cette conversion est très rapide. Aucun signe de neuropathie retardée induite par les organophosphorés n'a été relevé.

Le parathion-méthyl technique n'a aucun effet irritant sur l'oeil ni la peau.

Lors d'études de toxicité à court terme utilisant diverses voies d'administration et portant sur des rats, des chiens et des lapins, on a observé une inhibition de la cholinestérase du plasma, des érythrocytes et du cerveau ainsi qu'un certain nombre de signes liés aux effets cholinergiques. Lors d'une étude d'alimentation de 12 semaines sur des chiens, on a obtenu, pour la dose sans effet nocif observable, une valeur de 5 mg/kg de nourriture (soit l'équivalent de 0,1 mg/kg de poids corporel par jour). Lors d'une étude de toxicité par voie percutanée, effectuée pendant trois semaines sur des lapins, on a obtenu une dose sans effet observable de 10 mg/kg de poids corporel par jour. Lorsque les animaux étaient exposés par la voie respiratoire pendant trois semaines, la dose sans effet observable était de 0,9 mg/m³ d'air. A la dose de 2,6 mg/m³, on n'a observé qu'une légère inhibition de la cholinestérase plasmatique.

Des études de cancérogénicité et de toxicité à long terme ont été effectuées sur des souris et des rats. Pour les rats, la dose sans effet observable basée sur l'inhibition de la cholinestérase était de 0,1 mg/kg de poids corporel par jour. Les résultats de ces études n'ont fait ressortir aucun signe de cancérogénicité, ni chez les souris ni chez les rats. Dans une autre étude de deux ans effectuée sur des rats, on a toutefois relevé les signes d'un effet neurotoxique périphérique à la dose de 50 mg/kg de nourriture.

Le parathion-méthyl serait capable de provoquer l'alkylation de l'ADN *in vitro*. La plupart des études de génotoxicité *in vitro* portant sur des cellules bactériennes et mammaliennes ont donné des résultats positifs, alors que six études *in vivo* portant sur trois systèmes d'épreuve différents ont donné des résultats ambigus.

Les études portant sur la reproduction avec administration de doses toxiques (inhibition de la cholinestérase) n'ont pas produit d'effets systématiques sur la taille des portées et leur nombre, le taux de survie des petits ni la lactation. Aucun effet tératogène ou embryotoxique direct n'a été observé.

1.5 Effets sur l'homme

Plusieurs cas d'intoxication aiguë par le parathion-méthyl ont été signalés. Les symptômes sont caractéristiques d'une intoxication générale par les anticholinestérasiques organophosphorés. Il s'agit d'effets nerveux cholinergiques au niveau périphérique et au niveau central qui apparaissent dans les minutes qui suivent l'exposition. En cas d'exposition par voie percutanée, les symptômes peuvent s'aggraver pendant plus d'une journée et durer plusieurs jours.

Des études sur des volontaires soumis à des expositions répétées de longue durée ont montré que l'activité cholinestérasique du sang diminuait sans provoquer de manifestations cliniques.

Aucun cas de neuropathie périphérique retardée induite par les organophosphorés n'a été signalé. Dans un certain nombre de cas d'exposition multiple à des pesticides et notamment à du parathion-méthyl, on a observé des séquelles neurospsychiatriques.

Une augmentation du nombre des aberrations chromosomiques a été signalée dans des cas d'intoxication aiguë.

On ne possède aucune donnée obtenue sur l'homme qui puisse permettre d'évaluer les effets tératogènes du parathion-méthyl ou ses effets sur la reproduction.

Les études épidémiologiques disponibles sont consacrées à des expositions multiples aux pesticides et il n'est pas possible d'en déduire les effets qu'une exposition de longue durée au parathion-méthyl pourrait entraîner.

2 Conclusions

Le parathion-méthyl est un insecticide organophosphoré très toxique. Une exposition excessive due à la manipulation de ce produit au cours de la production, de l'utilisation ou par suite d'ingestion accidentelle ou intentionnelle peut entraîner une intoxication grave voire mortelle. Certaines formulations de parathion-méthyl peuvent, selon le cas, entraîner une irritation des yeux ou de la peau mais de toute façon, elles sont toutes facilement absorbées. On peut donc être dangereusement exposé à cet insecticide sans s'en rendre compte.

Le parathion-méthyl ne subsiste pas dans l'environnement. Il ne subit pas de bioconcentration et ne se transmet pas le long de la chaîne alimentaire. Il est rapidement décomposé par un grand nombre de microorganismes et autres éléments de la faune sauvage. Cet insecticide peut provoquer des dégâts dans les écosystèmes, mais seulement en cas d'exposition excessive dues à une utilisation défectueuse ou à des déversements accidentelles. Toutefois les insectes utiles et notamment les insectes pollinisateurs peuvent souffrir des épandages de parathion-méthyl.

C'est principalement par l'intermédiaire des denrées alimentaires que la population générale peut être exposée à des résidus de parathion-méthyl. Si l'on respecte les règles de bonne pratique agricole, il n'y a pas de raison que la dose journalière admissible fixée par le Comité d'experts FAO/OMS soit dépassée (0-0,02 mg/kg de poids corporel)). Il peut également y avoir exposition par voie percutanée lors de contacts accidentels avec des résidus foliaires dans des champs traités ou des zones voisines contaminées par des embruns de pesticides.

Moyennant de bonnes méthodes de travail et des précautions suffisantes en matières d'hygiène et de sécurité, le parathion-méthyl de devrait pas présenter de danger pour ceux qui lui sont exposés de par leur profession.

3 Recommandations

- Afin de protéger la santé et le bien-être des travailleurs et de la population générale il ne faut confier la manipulation et l'épandage du parathion-méthyl qu'à des personnes bien encadrées et bien formées qui utiliseront l'insecticide en prenant les mesures de sécurité nécessaires et se conformeront aux règles de bonne pratique en la matière.

- La fabrication, la formulation, l'utilisation agricole et l'élimination du parathion-méthyl doivent être conduites avec soin afin de réduire au minimum la contamination de l'environnement.

- Les travailleurs qui sont régulièrement exposés au parathion-méthyl doivent bénéficier d'un suivi médical approprié.

- Afin de réduire les risques pour l'ensemble de la population, il est recommandé de ne pas revenir sur une zone traitée avant 48 heures.

- Les autorités nationales devront fixer les délais pour les épandages avant récolte et les faire respecter.

- En raison de la forte toxicité du parathion-méthyl, cet insecticide ne doit pas être épandu à très bas volume à l'aide de dispositifs à main.

- Ne pas pulvériser sur les étendues d'eau. Choisir les horaires de manière à éviter de détruire les insectes pollinisateurs.

- Les données sur l'état de santé des travailleurs exposés uniquement au parathion-méthyl (c'est-à-dire employés à la fabrication et à la formulation de cet insecticide) devront être publiées afin que l'on puisse mieux en évaluer les risques pour la santé humaine.

- Des études à caractère plus définitif devront être menées sur les résidus de parathion-méthyl dans les denrées alimentaires fraîches.

- Il faudrait procéder à une évaluation plus concluante de la génotoxicité du parathion-méthyl.

RESUMEN Y EVALUACION, CONCLUSIONES Y RECOMENDACIONES

1 Resumen y evaluación

1.1 Exposición

El metilparatión es un insecticida organofosforado que se sintetizó por primera vez en la década de 1940. Es relativamente insoluble en agua, poco soluble en éter de petróleo y aceites minerales y fácilmente soluble en la mayoría de los disolventes orgánicos. El metilparatión puro se encuentra en forma de cristales blancos; el de calidad técnica tiene un color tostado claro y olor parecido al del ajo. Es térmicamente inestable y se descompone con rapidez a un pH superior a 8.

El método más común para la determinación del metilparatión es la cromatografía de gases con un detector de ionización de llama en álcali o bien con uno fotométrico de llama. Los límites de detección en el agua oscilan entre 0,01 y 0,1 μg/litro, y en el aire entre 0,1 y 1 ng/m^3. También son útiles como métodos de detección la cromatografía líquida de alta resolución y la cromatografía en capa fina.

En la distribución del metilparatión en el aire, el agua, el suelo y los organismos del medio ambiente influyen varios factores físicos, químicos y biológicos.

Los estudios realizados utilizando modelos de ecosistemas y la elaboración de modelos matemáticos indican que en el medio ambiente el metilparatión se reparte principalmente entre el aire y el suelo, con cantidades menores en las plantas y los animales. Prácticamente no hay desplazamiento a través del suelo, y ni el compuesto original ni los productos derivados de su degradación llegan normalmente al agua subterránea. El metilparatión presente en el aire procede sobre todo del rociado del compuesto, aunque se produce cierta volatilización con la evaporación del agua de las hojas y de la superficie del suelo. Los niveles habituales de metilparatión en la atmósfera en las zonas agrícolas oscilan entre una cantidad no detectable y unos 70 ng/m^3. Se ha observado que las concentraciones en el aire después del rociado disminuyen con rapidez en tres días, alcanzando los niveles habituales en unos nueve días. La concentración en el agua fluvial (en estudios de laboratorio) descendió al 80% de la inicial después de una hora, y transcurrida una semana era del 10%. El metilparatión se mantiene en el suelo más tiempo que en el aire o el agua, aunque en la retención influye mucho el tipo de suelo; el arenoso pierde los residuos del compuesto con mayor rapidez que las margas. Los residuos de la superficie de las plantas y del interior de las hojas disminuyen rápidamente, con una semivida del orden de unas horas; el metilparatión desaparece totalmente en unos 6-7 días.

Los animales pueden degradar el metilparatión y eliminar los productos de degradación en un período muy breve de tiempo. El proceso es más lento en los vertebrados inferiores y en los invertebrados que en los mamíferos y las aves. Los factores de bioconcentración son bajos y los niveles acumulados de metilparatión transitorios.

La descomposición microbiana es con diferencia el mecanismo más importante de degradación del metilparatión en el medio ambiente. La desaparición del compuesto en el campo y en ecosistemas utilizados como modelo es más rápida de lo que habían permitido suponer los estudios de laboratodeloorio. Esto se debe a la variedad de microorganismos que son capaces de degradarlo en distintos hábitats y circunstancias. La presencia de sedimentos o de superficies de plantas, que aumenta la población microbiana, acelera el ritmo de degradación del metilparatión.

El metilparatión puede sufrir degradación oxidativa por acción de la radiación ultravioleta o la luz solar, convirtiéndose en metilparaoxón, que es menos estable; las películas de rociado se degradan por acción de la radiación ultravioleta con una semivida aproximada de 40 horas. Sin embargo, se ha estimado que la contribución de la fotolisis a la desaparición total en un sistema acuático es sólo de un 4%. También se produce hidrólisis del metilparatión en condiciones alcalinas, en las que es más rápida. La salinidad elevada favorece asimismo la hidrólisis del compuesto. En sedimentos muy reductores se registraron semividas de unos minutos, aunque el metilparatión es más estable cuando está adsorbido sobre otros sedimentos.

En las ciudades situadas en el centro de las zonas agrícolas de los Estados Unidos, las concentraciones de metilparatión en el aire variaban con las estaciones y alcanzaban el punto más alto en agosto o septiembre; los niveles máximos registrados durante los estudios fueron fundamentalmente del orden de 100-800 ng/m^3 durante el período vegetativo. Las concentraciones en el agua natural de las zonas agrícolas de los Estados Unidos llegaron a 0,46 μg/litro, con los niveles más altos en el verano. Son muy pocos los informes publicados en todo el mundo sobre los residuos de metilparatión en los alimentos. En los Estados Unidos, se han notificado en general niveles muy bajos de residuos de metilparatión en los productos alimenticios, con un pequeño número de muestras aisladas por encima de los límites máximos de residuos (LMR). En todos los estudios publicados sobre la alimentación sólo se detectaron niveles ínfimos de metilparatión. En las encuestas sobre la cesta de la compra realizadas en los Estados Unidos entre 1966 y 1969, las cantidades mayores de residuos de metilparatión se encontraron en las hortalizas de hoja (hasta 2 mg/kg) y en las de raíz (hasta 1 mg/kg). En la preparación, cocción y almacenamiento de los alimentos se descomponen los residuos de metilparatión, reduciéndose ulteriormente la exposición humana. Las frutas y hortalizas sin elaborar pueden contener más residuos después de un uso indebido del producto.

La producción, formulación, manipulación y uso del metilparatión como insecticida pueden ser, en principio, fuente de exposición para las personas. Las principales vías de exposición de los trabajadores son el contacto cutáneo y, en menor medida, la inhalación.

En un estudio sobre personas encargadas del rociado en fincas (trabajadores no protegidos que utilizaban rociadores manuales de volumen ultrabajo), a partir del *p*-nitrofenol excretado en la orina se calculó una ingestión de 0,4-13 mg de metilparatión cada 24 horas. También se puede sufrir exposición si se entra en los cultivos demasiado pronto después de tratarlos.

La población general puede estar expuesta a residuos de metilparatión presentes en el aire, el agua y los alimentos como consecuencia de prácticas agrícolas y forestales con un uso indebido del producto, que provoca la contaminación de los campos, los cultivos, el agua y el aire debido al rociado parcial fuera del objetivo.

1.2 Ingestión, metabolismo y excreción

El metilparatión se absorbe fácilmente por todas las vías de exposición (oral, cutánea, respiratoria) y se distribuye con rapidez por los tejidos del cuerpo. Se detectaron concentraciones máximas en diversos órganos 1-2 horas después del tratamiento. Después de la administración, la transformación del metilparatión en metilparaoxón se produce en unos minutos. En perros se determinó una semivida terminal media de 7,2 horas tras la administración intravenosa de metilparatión. El hígado es el principal órgano de metabolización y desintoxicación. El metilparatión o el metilparaoxón se destoxifican en el hígado sobre todo mediante oxidación, hidrólisis y desmetilación o desarilación con glutatión reducido. Los productos de la reacción son el *O*-metil *O-p*-nitrofenilfosfotioato, o bien los ácidos dimetilfosfotioico o dimetilfosfórico, y el *p*-nitrofenol. Por consiguiente, se puede estimar la exposición midiendo la excreción urinaria de *p*-nitrofenol; en voluntarios humanos fue del 60% en cuatro horas y prácticamente del 100% en 24 horas. El metabolismo del metilparatión es importante para la toxicidad específica selectiva y la aparición de resistencia. Le eliminación de esta sustancia y sus productos derivados tiene lugar primordialmente por la orina. En estudios realizados en ratones con ^{32}P-metilparatión (marcado radiactivamente) se observó un 75% de radiactividad en la orina y hasta un 10% en las heces después de 72 horas.

1.3 Efectos en los seres vivos del medio ambiente

Los microorganismos pueden utilizar el metilparatión como fuente de carbono, y en el estudio de una comunidad natural se vio que concentraciones de hasta 5 mg/litro aumentaban la biomasa y la actividad reproductora. En las bacterias y los actinomicetos se observó un efecto positivo del metilparatión, mientras que los hongos y las levaduras tenían menor capacidad para utilizar la sustancia. Con una concentración aproximada de 5 mg/litro se produjo una inhibición del 50% del crecimiento de una diatomea. Concentraciones de metilparatión comprendidas entre 25 y 80 μg/litro redujeron el crecimiento celular de las algas clorofíceas unicelulares. Las poblaciones de algas adquirieron tolerancia tras varias semanas de exposición.

El metilparatión es muy tóxico para los invertebrados acuáticos, oscilando casi siempre la CL_{50} entre < 1 μg y alrededor de 40 μg/litro. Hay un pequeño número de especies de artrópodos que son menos susceptibles. El nivel sin efecto para *Daphnia magna* es de 1,2 μg/litro. Los moluscos son mucho menos susceptibles, con CL_{50} entre 12 y 25 mg/litro.

La mayoría de las especies de peces, tanto de agua dulce como de mar, tienen una CL_{50} de 6 a 25 mg/litro, pero hay un pequeño número de especies cuya sensibilidad al metilparatión es considerablemente mayor o menor. La toxicidad aguda para los anfibios es análoga a la de los peces.

Se han observado los efectos sobre poblaciones en las comunidades de invertebrados acuáticos de estanques experimentales tratados con metilparatión. Las concentraciones necesarias para producir esos efectos se alcanzarían sólo con un rociado excesivo de las masas de agua, e incluso en este caso durarían muy poco tiempo. Por consiguiente, en condiciones normales no es probable que se observen efectos sobre las poblaciones. Tampoco los es que la acción letal sobre los invertebrados acuáticos provoque efectos duraderos.

Hay que tener cuidado para evitar un rociado excesivo de estanques, ríos y lagos al utilizar el metilparatión. Nunca se debe efectuar la operación con viento.

El metilparatión es un insecticida no selectivo que mata especies beneficiosas tan fácilmente como las plagas. Se ha notificado la muerte de abejas después de su aplicación. Sus efectos sobre esta especie fueron más graves que los de otros insecticidas. Las abejas africanizadas son más tolerantes al metilparatión que las razas europeas.

El metilparatión fue moderadamente tóxico para las aves en estudios de laboratorio, con una DL_{50} oral aguda comprendida entre 3 y 8 mg/kg de peso corporal. La CL_{50} en la dieta osciló entre 70 y 680 mg/kg de alimentos. No hay indicios de que las aves puedan verse afectadas negativamente con la utilización recomendada en el campo.

Hay que tener el máximo cuidado al programar el rociado con metilparatión, a fin de evitar los efectos adversos sobre las abejas.

1.4 Efectos en los animales de experimentación y en sistemas de prueba in vitro

Los valores de la DL_{50} del metilparatión por vía oral en roedores oscilan entre 3 y 35 mg/kg de peso corporal, y los valores por vía cutánea entre 44 y 67 mg/kg de peso corporal.

El envenenamiento por metilparatión provoca los signos colinérgicos habituales de los organofosfatos, atribuidos a la acumulación de acetilcolina en la terminaciones nerviosas. El metilparatión adquiere la toxicidad al metabolizarse a metilparaoxón, en un proceso que es muy rápido. No se han observado indicios de neuropatía retardada inducida por compuestos organofosforados.

Se ha comprobado que el metilparatión de calidad técnica no tiene potencial de irritación primaria de los ojos o la piel.

En estudios de toxicidad de corta duración, utilizando diversas vías de administración en ratas, perros y conejos, se observó inhibición de la colinesterasa del plasma, los eritrocitos y el cerebro, así como signos colinérgicos conexos. En un estudio de alimentación durante 12 semanas con perros, el nivel sin efectos adversos observados (NOAEL) fue de 5 mg/kg de la dieta (equivalente a 0,1 mg/kg de peso corporal al día). En un estudio de toxicidad cutánea de tres semanas en conejos, el nivel sin efectos observados (NOEL) fue de 10 mg/kg de peso corporal al día. La exposición por inhalación durante tres semanas dio como resultado un NOEL de 0,9 mg/m^3 de aire. Con 2,6 mg/m^3 solamente se observó una ligera inhibición de la colinesterasa del plasma.

Se realizaron estudios de toxicidad/teratogenicidad de larga duración con ratones y ratas. El NOEL para las ratas fue de 0,1 mg/kg de peso corporal al día, basado en la inhibición de la colinesterasa. No hay pruebas de carcinogenicidad en ratones y ratas tras una exposición de larga duración. Sin embargo, en otro estudio de dos años con ratas se detectó un efecto neurotóxico periférico con una dosis de 50 mg/kg de la dieta.

Se ha informado que el metilparatión tiene propiedades alquilizantes del ADN *in vitro*. Los resultados de la mayoría de los estudios de genotoxicidad *in vitro* con células tanto bacterianas como de mamífero fueron positivos, mientras que en seis estudios *in vivo*, utilizando tres sistemas de prueba distintos, los resultados fueron equívocos.

En estudios de reproducción con niveles de dosificación tóxicos (inhibición de la colinesterasa), no se observaron efectos constantes sobre el tamaño de la camada, el número de partos, la tasa de supervivencia de las crías y el rendimiento de la lactación. No se detectó ningún efecto teratogénico o embriotóxico primario.

1.5 Efectos en la especie humana

Se han registrado varios casos de intoxicación aguda por metilparatión. Los signos y síntomas son los característicos de la intoxicación sistémica por compuestos organofosforados inhibidores de la colinesterasa. Cabe mencionar entre ellos las manifestaciones del sistema nervioso colinérgico periférico y central, que aparecen apenas unos minutos después de la exposición. En el caso de la exposición cutánea, la gravedad de los síntomas puede ir en aumento durante más de un día y pueden durar varios días.

Los estudios con voluntarios sometidos a exposiciones repetidas de larga duración parecen indicar que hay una disminución de la actividad de la colinesterasa de la sangre, sin manifestaciones clínicas.

No se ha informado de ningún caso de neuropatía periférica retardada inducida por compuestos organofosforados. Se han descrito secuelas neuropsiquiátricas en casos de exposición múltiple a plaguicidas, entre ellos el metilparatión.

En casos de intoxicaciones agudas, se ha detectado un aumento de las aberraciones cromosómicas.

No se dispone de datos relativos al metilparatión en la especie humana que permitan evaluar los efectos teratogénicos y sobre la reproducción.

Los estudios epidemiológicos disponibles se refieren a una exposición múltiple a plaguicidas, y no es posible evaluar los efectos de una exposición de larga duración al metilparatión.

2 Conclusiones

El metilparatión es un éster organofosfórico muy tóxico, utilizado como insecticida. Una exposición excesiva al manejarlo durante su fabricación y uso o por ingestión accidental o intencionada puede ocasionar una intoxicación grave o letal. Las formulaciones de metilparatión unas veces son irritantes y otras no para los ojos o la piel, pero se absorben fácilmente. Por consiguiente, pueden producirse exposiciones peligrosas sin advertirlo.

El metilparatión no se mantiene mucho tiempo en el medio ambiente, no se produce bioconcentración y no se desplaza a través de la cadena alimentaria. Lo degradan con rapidez numerosos microorganismos y otros tipos de seres vivos presentes en el medio ambiente. Este insecticida puede ocasionar daños a ecosistemas solamente en casos de una exposición muy intensa causada por el uso indebido o escapes accidentales; sin embargo, el rociado con metilparatión representa un riesgo para los insectos polinizadores y otros que son beneficiosos.

La exposición de la población general a los residuos del metilparatión tiene lugar fundamentalmente por medio de los alimentos. Si se siguen buenas prácticas agrícolas, no se supera la ingesta diaria admisible (0-0,02 mg/kg de peso corporal) establecida por la FAO/OMS. Puede haber exposición cutánea accidental por contacto con residuos foliares en campos rociados o en zonas adyacentes a los lugares que se están rociando, como consecuencia de pérdidas del producto que no llegan a su objetivo.

Con buenas prácticas de trabajo, medidas higiénicas y precauciones de seguridad, no es probable que el metilparatión represente un riesgo para las personas con exposición profesional.

3 Recomendaciones

- Para salvaguardar la salud y el bienestar de los trabajadores y de la población general, el manejo y la aplicación del metilparatión sólo se debería encomendar, bajo una atenta supervisión, a personas bien capacitadas que se ajusten a las medidas de seguridad adecuadas y utilicen el producto de acuerdo con las buenas prácticas de aplicación.

- Se debe prestar particular atención a la fabricación, la formulación, el uso agrícola y la eliminación del metilparatión, a fin de reducir al mínimo la contaminación del medio ambiente.

- Los trabajadores regularmente expuestos deberían ser objeto de vigilancia y exámenes médicos adecuados.

- A fin de reducir al mínimo el riesgo para todas las personas, se recomienda esperar 48 horas después del rociado antes de entrar de nuevo en cualquier zona tratada.

- Las autoridades nacionales deberían establecer intervalos sin tratamiento antes de la recolección y obligar a respetarlos.

- A la vista de la elevada toxicidad del metilparatión, se debe excluir este producto de la aplicación mediante rociado de volumen ultrabajo aplicado manualmente.

- No se han de rociar masas de agua. Hay que elegir los momentos de la aplicación de manera que se evite la muerte de insectos polinizadores.

- Se debe hacer pública la información relativa al estado de salud de los trabajadores expuestos exclusivamente al metilparatión (es decir, en la fabricación, la formulación), con objeto de evaluar mejor los riesgos de este producto químico para la salud humana.

- Deberían llevarse a cabo estudios más definitivos sobre los residuos de metilparatión en los alimentos frescos.

- Debería realizarse una evaluación genotóxica más definitiva del metilparatión.

www.ingramcontent.com/pod-product-compliance
Lightning Source LLC
Chambersburg PA
CBHW071159210326
41597CB00016B/1600